Encyclopedia of the Animal World

MAMMALS

The Small Plant-Eaters

Martyn Bramwell & Steve Parker

WITH CONTRIBUTIONS BY

Jill Bailey and Linda Losito

Facts On File

New York • Oxford

THE SMALL PLANT-EATERS
The Encyclopedia of the Animal World: Mammals

Managing Editor: Lionel Bender
Art Editor: Ben White
Designer: Malcolm Smythe
Text Editor: Miles Litvinoff
Assistant Editor: Madeleine Samuel
Project Editor: Graham Bateman
Production: Clive Sparling

Media conversion and typesetting: Peter MacDonald and Una Macnamara

Planned and produced by:
Equinox (Oxford) Limited,
Musterlin House, Jordan Hill Road,
Oxford OX2 8DP

Prepared by Lionheart Books

Library of Congress
Cataloging-in-Publication Data
Bramwell, Martyn.
Mammals: the small plant-eaters/Martyn Bramwell and Steve Parker, with contributions by Jill Bailey and Linda Losito.
p.96,cm.22.5×27.5 (The Encyclopedia of the animal world)
Bibliography: p.1
Includes index.
Summary: Examines those mammals which are herbivores, including koalas, marmots, hamsters, and squirrels.
1. Herbivores – Juvenile literature. 2. Mammals – Juvenile literature [1. Herbivores. 2. Mammals.] I. Parker, Steve. II. Title. III. Series.

QL706.2.B72 1988 599-dc19
88-16934

ISBN 0-8160-1958-4

Published in North America by
Facts on File, Inc.,
460 Park Avenue South,
New York, N.Y. 10016

Origination by Alpha Reprographics Ltd, Harefield, Middx, England

Printed in Italy.

10 9 8 7 6 5 4 3 2 1

FACT PANEL: Key to symbols denoting general features of animals

SYMBOLS WITH NO WORDS

Activity time

- Nocturnal
- Daytime
- Dawn/Dusk
- All the time

Group size

- Solitary
- Pairs
- Small groups (up to 10)
- Herds/Flocks
- Variable

Conservation status

- All species threatened
- Some species threatened
- No species threatened (no symbol)

SYMBOLS NEXT TO HEADINGS

Habitat

- General
- Mountain/Moorland
- Desert
- Sea
- Amphibious
- Tundra
- Forest/Woodland
- Grassland
- Freshwater

Diet

- Other animals
- Plants
- Animals and Plants

Breeding

- Seasonal (at fixed times)
- Non-seasonal (at any time)

CONTENTS

PREFACE

The National Wildlife Federation

For the wildlife of the world, 1936 was a very big year. That's when the National Wildlife Federation formed to help conserve the millions of species of animals and plants that call Earth their home. In trying to do such an important job, the Federation has grown to be the largest conservation group of its kind.

Today, plants and animals face more dangers than ever before. As the human population grows and takes over more and more land, the wild places of the world disappear. As people produce more and more chemicals and cars and other products to make life better for themselves, the environment often becomes worse for wildlife.

But there is some good news. Many animals are better off today than when the National Wildlife Federation began. Alligators, wild turkeys, deer, wood ducks, and others are thriving – thanks to the hard work of everyone who cares about wildlife.

The Federation's number one job has always been education. We teach kids the wonders of nature through *Your Big Backyard* and *Ranger Rick* magazines and our annual National Wildlife Week celebration. We teach grown-ups the importance of a clean environment through *National Wildlife* and *International Wildlife* magazines. And we help teachers teach about wildlife with our environmental education activity series called *Naturescope.*

The National Wildlife Federation is nearly five million people, all working as one. We all know that by helping wildlife, we are also helping ourselves. Together we have helped pass laws that have cleaned up our air and water, protected endangered species, and left grand old forests standing tall.

You can help too. Every time you plant a bush that becomes a home to a butterfly, every time you help clean a lake or river of trash, every time you walk instead of asking for a ride in a car – you are part of the wildlife team.

You are also doing your part by learning all you can about the wildlife of the world. That's why the National Wildlife Federation is happy to help bring you this Encyclopedia. We hope you enjoy it.

Jay D. Hair, President
National Wildlife Federation

INTRODUCTION

The Encyclopedia of the Animal World surveys the main groups and species of animals alive today. Written by a team of specialists, it includes the most current information and the newest ideas on animal behavior and survival. The Encyclopedia looks at how the shape and form of an animal reflect its life-style – the ways in which a creature's size, color, feeding methods and defenses have all evolved in relationship to a particular diet, climate and habitat. Discussed also are the ways in which human activities often disrupt natural ecosystems and threaten the survival of many species.

In this Encyclopedia the animals are grouped on the basis of their body structure and their evolution from common ancestors. Thus, there are single volumes or groups of volumes on mammals, birds, reptiles and amphibians, fish, insects and so on. Within these major categories, the animals are grouped according to their feeding habits or general life-styles. Because there is so much information on the animals in two of these major categories, there are four volumes devoted to mammals (*The Small Plant-Eaters; The Hunters; The Large Plant-Eaters; Primates, Insect-Eaters and Baleen Whales*) and three to birds (*Waterbirds; Aerial Hunters and Flightless Birds; Plant- and Seed-Eaters*).

This volume, *Mammals – The Small Plant-Eaters*, includes entries on rats, mice, squirrels, rabbits, hares, hamsters, porcupines, sloths, wombats and the koala. Together they number some 1,700 species. Compared to mammals like elephants and rhinoceroses, the majority of these animals are *small*. Nevertheless, their life-styles are intriguing. Some, such as voles, burrow. Others, such as tree squirrels, climb. Some species of rats and mice live close to people. All are *herbivores*, which means they eat plant material as at least the main part of their diet. But their diets vary greatly. Beavers feed on wood and gerbils on seeds, while the koala eats only eucalyptus leaves and the Honey possum feeds solely on nectar.

The most important group of small herbivores are the rodents – rats, mice, voles, gerbils, squirrels etc. They are a very successful group of mammals, occupying nearly every type of land habitat. Their success lies with their ability both to adapt easily to changing conditions and to breed quickly and in great numbers. Almost as important, but quite unrelated to the rodents, are the lagomorphs (hare-like animals), the most familiar of which are the rabbits and hares, but also including the pikas.

Other small plant-eaters included here are the hyraxes, sloths and colugos, each of which are quite distinct groups of animals, with different evolutionary histories.

The marsupial equivalents of some of these animals included here are the koala, phalangers, wombats and the Honey possum.

Also included in this volume are introductions to the animal kingdom and to mammals in general.

Each article in this Encyclopedia is devoted to an individual species or group of closely related species. The text starts with a short scene-setting story that highlights one or more of the animal's unique features. It then continues with details of the most interesting aspects of the animal's physical features and abilities, diet and feeding behavior, and general life-style. It also covers conservation and the animal's relationships with people.

A fact panel provides easy reference to the main features of distribution (natural, not introductions to other areas by humans), habitat, diet, size, color, pregnancy and birth, and lifespan. (An explanation of the color coded symbols is given on page 2 of the book.) The panel also includes a list of the common and scientific (Latin) names of species mentioned in the main text and photo captions. For species illustrated in major artwork panels but not described elsewhere, the names are given in the caption accompanying the artwork. In such illustrations, all animals are shown to scale; actual dimensions may be found in the text. To help the reader appreciate the size of the animals, in the upper right part of the page at the beginning of an article are scale drawings comparing the size of the species with that of a human being (or of a human foot).

Many species of animal are threatened with extinction as a result of human activities. In this Encyclopedia the following terms are used to show the status of a species as defined by the International Union for the Conservation of Nature and Natural Resources:

Endangered – in danger of extinction unless their habitat is no longer destroyed and they are not hunted by people.

Vulnerable – likely to become endangered in the near future.

Rare – exist in small numbers but neither endangered nor vulnerable at present.

A glossary provides definitions of technical terms used in the book. A common name and scientific (Latin) name index provide easy access to text and illustrations.

LIVING THINGS

Living things come in a huge variety of shapes and sizes. The largest animal on Earth, the Blue whale, grows to over 110ft long and may weigh more than 185 tons. By contrast, a spoonful of pond water may hold 1,000 animals and plants, each too small to see with the naked eye.

There are so many different kinds (species) of living things (organisms) on this planet that it is difficult to understand how they are related to each other. And so, scientists sort them into groups. The members of each group have a number of features in common. Each group can be divided into several smaller groups, whose members share even more common features, and so on.

The biggest groups of all are the kingdoms. There are five kingdoms – Animals, Plants, Fungi, Protozoa (one-celled organisms) and Monera (mainly bacteria).

THE ANIMAL KINGDOM

This contains at least 1,250,000 different kinds of living thing that range from birds and fish to insects and human beings. (Some scientists believe that there may be as many as 30 million kinds. Each year hundreds of new organisms are discovered.)

Animals feed on plants, on other animals, or on their dead remains. They cannot make their own food from non-organic materials as plants do from sunlight and materials taken in by the roots and leaves. Animals take food into their bodies, digest it and get rid of waste materials. Many animals can move about, but some, like sponges and barnacles, are fixed in one place.

LEVELS OF ORGANIZATION

The Animal Kingdom is first divided into groups according to body structure. There are several basic body plans, as shown opposite. Each is based on the cell as the basic building unit. A cell is a tiny bag of living material, surrounded by a thin skin or membrane. Tiny holes in the membrane allow materials to pass from one cell to another.

The simplest animals, such as sponges, have a body based on two layers of cells, but the cells are not linked or coordinated, and there are no muscle or nerve cells. Cells lining the body cavity take in food and digest it.

Animals like jellyfish and sea anemones also have a two-layered body wall, but they have well organized muscle and nerve systems. Food is partly digested by juices secreted into the body cavity, which acts like a simple gut.

Flatworms have a three-layered body, a distinct front and rear end, and their senses and nerve cells are concentrated at the front end.

More advanced animals, such as earthworms, have two openings to the gut – a mouth and an anus. This allows different parts of the gut to be specialized for different aspects of digestion, so a wider range of diets is possible. These animals also have a fluid-filled body cavity (coelom) in which different body organs can be suspended by sheets of tissue. This allows the body structure to become much more specialized and varied. All the more complex animals, including human beings, are based on this structural plan.

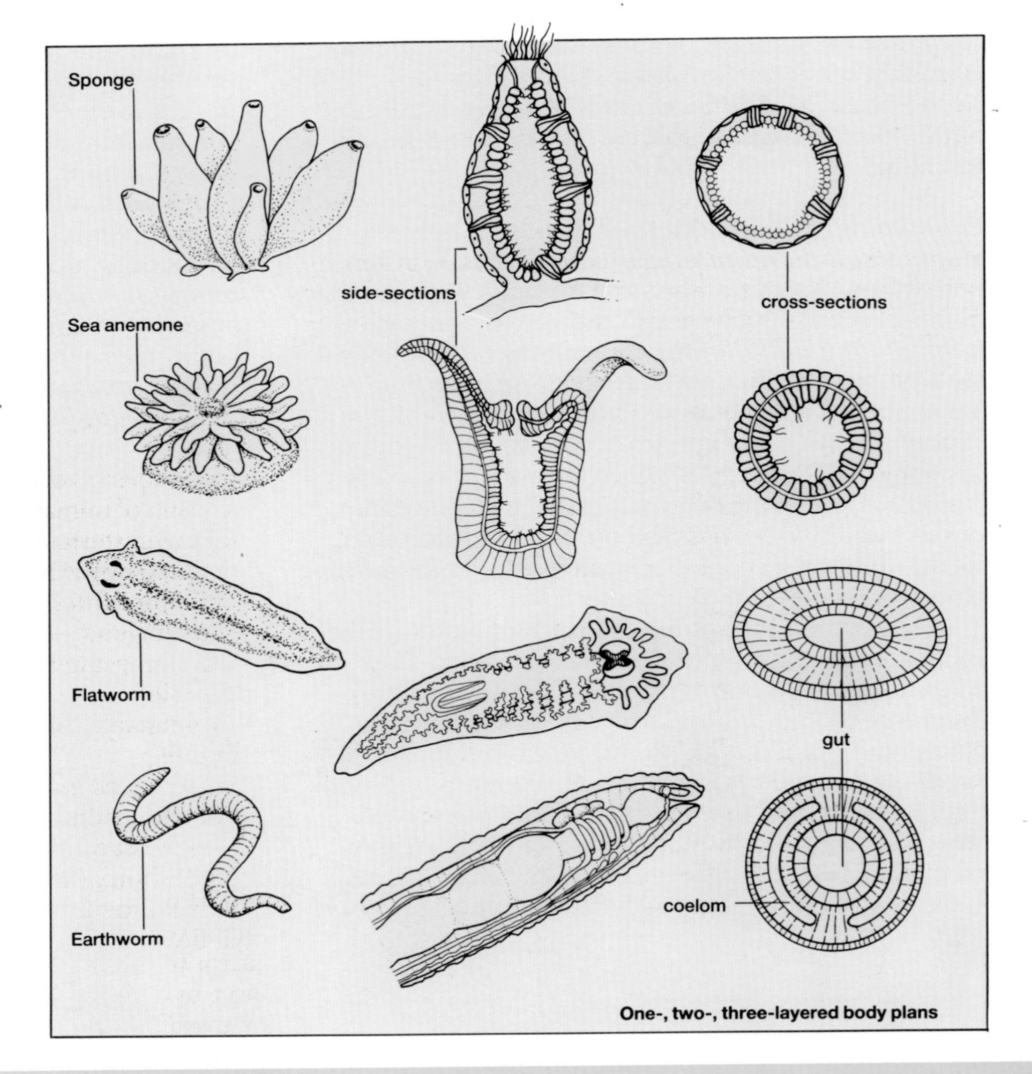

One-, two-, three-layered body plans

NAMING LIVING THINGS

All the different levels of grouping have their own names. Kingdoms are divided into phyla, which are divided into subphyla, classes, orders, families, genera and species. Members of the same species can mate and produce fertile offspring.

Each species has a double-barrelled scientific name. The first name is its genus and the second the name of that species. These names are in Latin, and are usually written in italics, with only the genus name starting with a capital letter. Often the same species is called by different everyday names in different countries. For example, the European elk is called a moose in North America, and the North American elk is the same species as the European Red deer and the Himalayan shou. By giving each species its own Latin name, people from different countries can be sure they are talking about the same kind of organism.

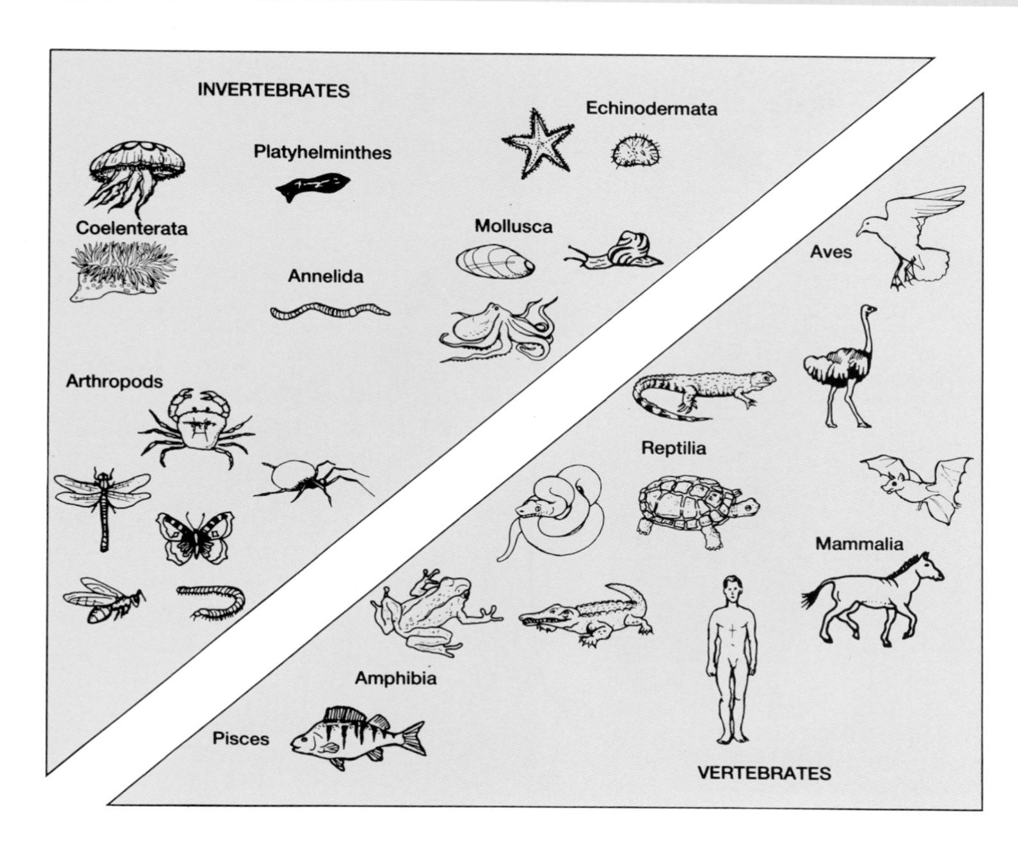

▲One of the most important divisions in the Animal Kingdom is between invertebrates (animals without backbones) and vertebrates (animals with backbones). These two groups can be further subdivided according to the arrangement of the different body parts, feeding habits, types of teeth, and so on.

▼By looking at the scientific categorization (classification) of the tiger, Latin name *Panthera tigris*, you can see how the grouping system works.

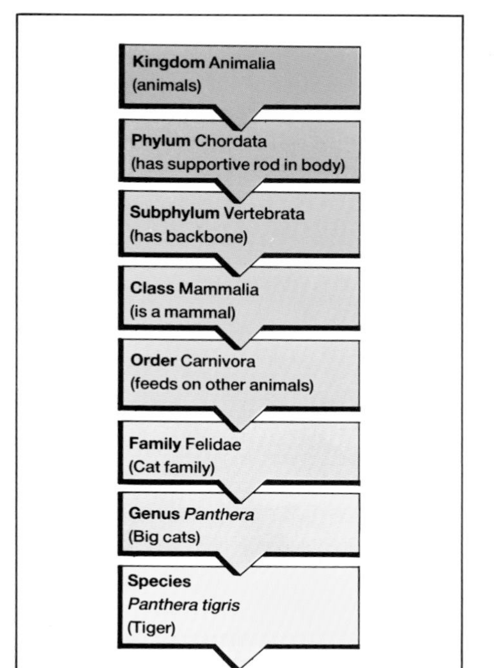

WHAT IS A MAMMAL?

The mammals are one of the most successful groups of animals. They are found in almost every habitat, from the Arctic tundra to the Sahara Desert, from the oceans to the mountain-tops. One group of mammals, the bats, has even taken to the air.

Mammals are vertebrates: they have an internal bony skeleton for support. The backbone is made up of a series of bony units called vertebrae. A bony shoulder girdle and pelvic girdle are attached to the backbone. A pair of forelegs is attached to the shoulder girdle and a pair of hind legs to the pelvic girdle (see illustration below).

Unlike fish, reptiles and amphibians (other vertebrate groups), mammals are able to keep their inside temperature fairly constant by producing the heat they need from their own body processes. Scientists now use the term endotherm to describe this type of animal rather than the more familiar term warm-blooded, which can be misleading. The blood of so-called cold-blooded animals, such as frogs and snakes, is cold if their surroundings are cold, but after sunbathing their blood may actually be hotter than the blood of warm-blooded mammals. Being able to control their body temperature regardless of the temperature of their surroundings has allowed the mammals to colonize most parts of the globe. Birds are also endotherms and mammals.

▲A Cape fox vixen suckles her cubs. The supply of milk helps to give the young a good start in life by providing the necessary food for growth.

The skeleton of a Grey wolf

Skull housing large brain
Shoulder girdle
Pelvic girdle
Backbone
Single bone of lower jaw
Rib-cage
5-digit feet

MAMMALS' SPECIAL FEATURES

The two main features that distinguish mammals from other vertebrates are hair and milk. Hairs trap a layer of air next to the skin. Air does not easily allow heat to pass through it, so this helps the mammal to stay warm.

When mammals reproduce, the female produces milk from special mammary glands and releases it through nipples on her belly. The milk forms a complete food for the baby mammal until it is strong enough to find food for itself. Milk production has allowed some mammals to give birth to young which may be small and helpless. These mammals can produce more young at a time than if the young had to be independent at birth. Because the young spend a long time with their parents while they grow up, this provides an opportunity to learn from experience and improves their chances of survival.

The mammal skull contains a large brain. The cerebral hemispheres of the brain (the parts dealing with consciousness, mental ability and intelligence) are very large.

The lower jaw is formed from a single bone, which makes it very strong. Mammal teeth are usually of various shapes and sizes, specialized for particular diets. A bony plate, the secondary palate, separates the nose passages from the mouth cavity, allowing a mammal to breathe when its mouth is full. This means that a mammal can spend as much time as it likes chewing its food before swallowing it. Most other vertebrates have to swallow their food whole, or they would run out of breath.

The mammal body is divided by a sheet of muscle, the diaphragm, at the base of the rib cage. Movements of the diaphragm help the animal breathe efficiently. Contraction of the diaphragm, along with moving of the ribs, sucks air into the lungs. The reverse process pushes air out of the lungs.

TYPES OF MAMMALS

There are three classes of mammals, distinguished by the way in which they breed (as shown below).

The Monotremes

These are the egg-laying mammals. They are the most primitive mammals. There are only three species: the duck-billed platypus and two kinds of echidna or spiny anteater. Monotremes are found only in Australia and New Guinea. They lay eggs with leathery shells, rather like those of reptiles. These hatch into tiny, little-developed young. Monotremes have no nipples. The young cling to the fur on their mother's belly and suck at the milk oozing out of the skin where the milk glands open to the surface.

The Marsupials

In this class, which includes the kangaroos, the koala and the opossums, the young develop inside the mother's womb, but are born at a very early stage of development. At birth, they look like tiny grubs. They climb up their mother's fur into a pouch on her belly. Inside the pouch are nipples which produce milk.

The Placental Mammals

This is the largest and most "advanced" group of mammals. The young develop inside their mother's womb attached to a placenta. This is a special structure which supplies them with food and oxygen from the mother's blood and carries away their waste products. This food supply allows the young to reach a more advanced stage before being born. Placental mammal mothers produce milk from nipples on their bellies.

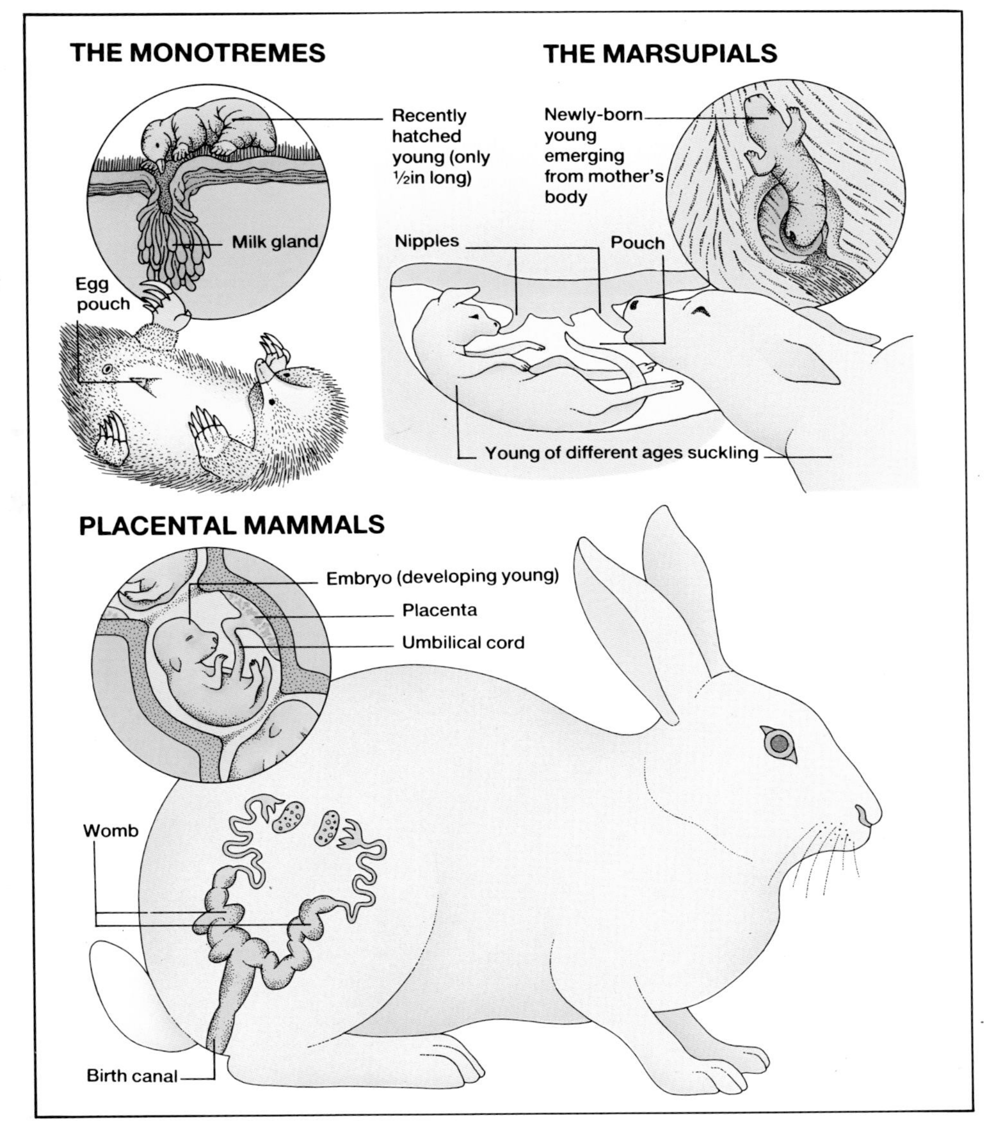

▼Some young mammals, like this new-born gazelle, can stand soon after birth.

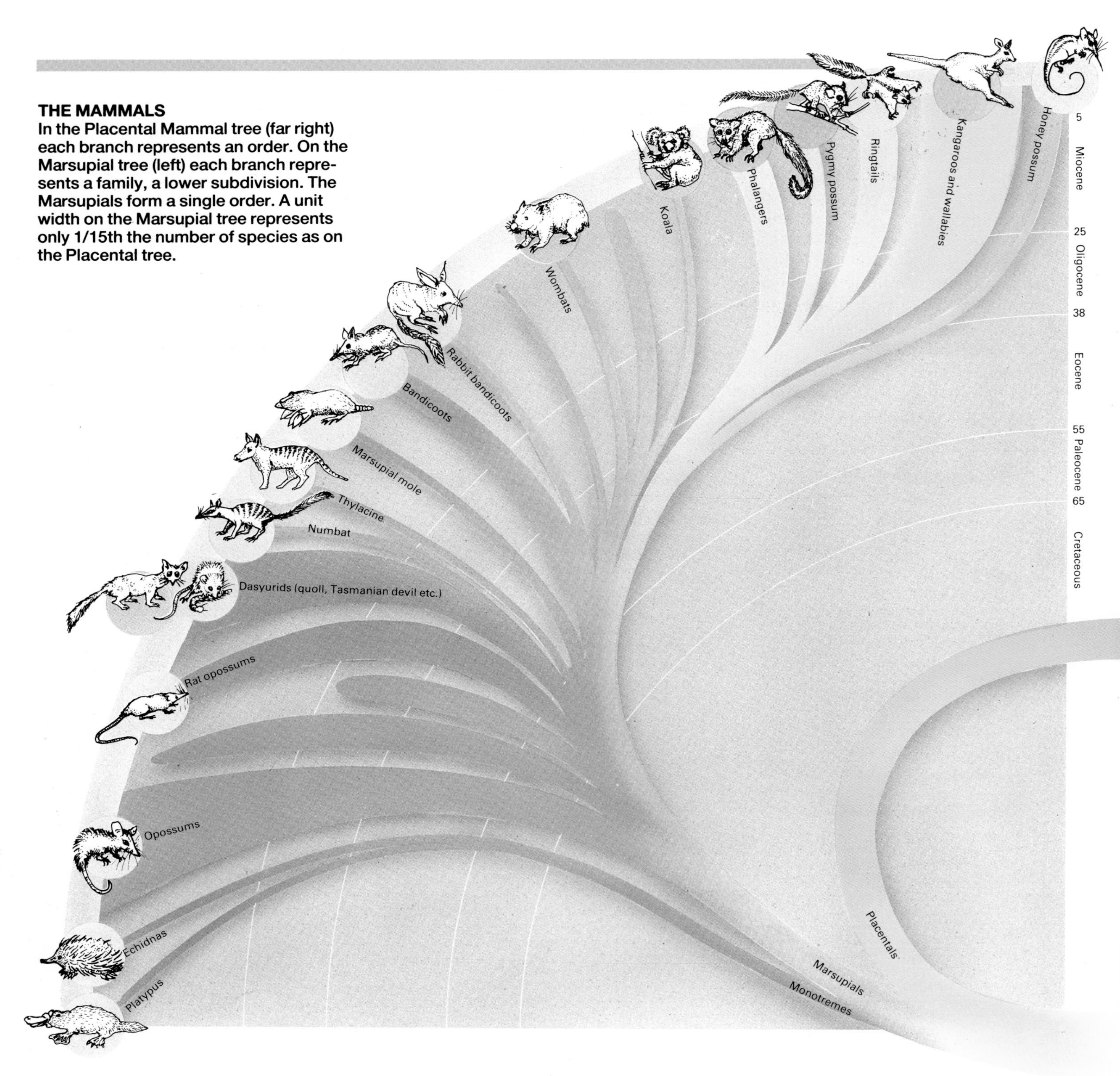

THE MAMMALS
In the Placental Mammal tree (far right) each branch represents an order. On the Marsupial tree (left) each branch represents a family, a lower subdivision. The Marsupials form a single order. A unit width on the Marsupial tree represents only 1/15th the number of species as on the Placental tree.

CLASSIFYING THE MAMMALS

There are over 4,000 different species of mammal, around 1,000 genera and 135 families. Mammals are grouped into Monotremes (3 species), Marsupials (266) and Placental mammals (over 3,750, of which a quarter are bats). They can be further sorted according to their body structure, especially the number, shapes and arrangement of the bones and teeth.

Each mammal's body is suited it to its particular way of life. Thus cheetahs have long legs and flexible backbones

▲This mammal family tree shows how the mammals have increased in variety with time. The ancestors of the mammals first appeared on Earth about 300 million years ago. Since then, mammals have evolved a wide range of feeding habits and life-styles, allowing them to spread to most parts of the globe.

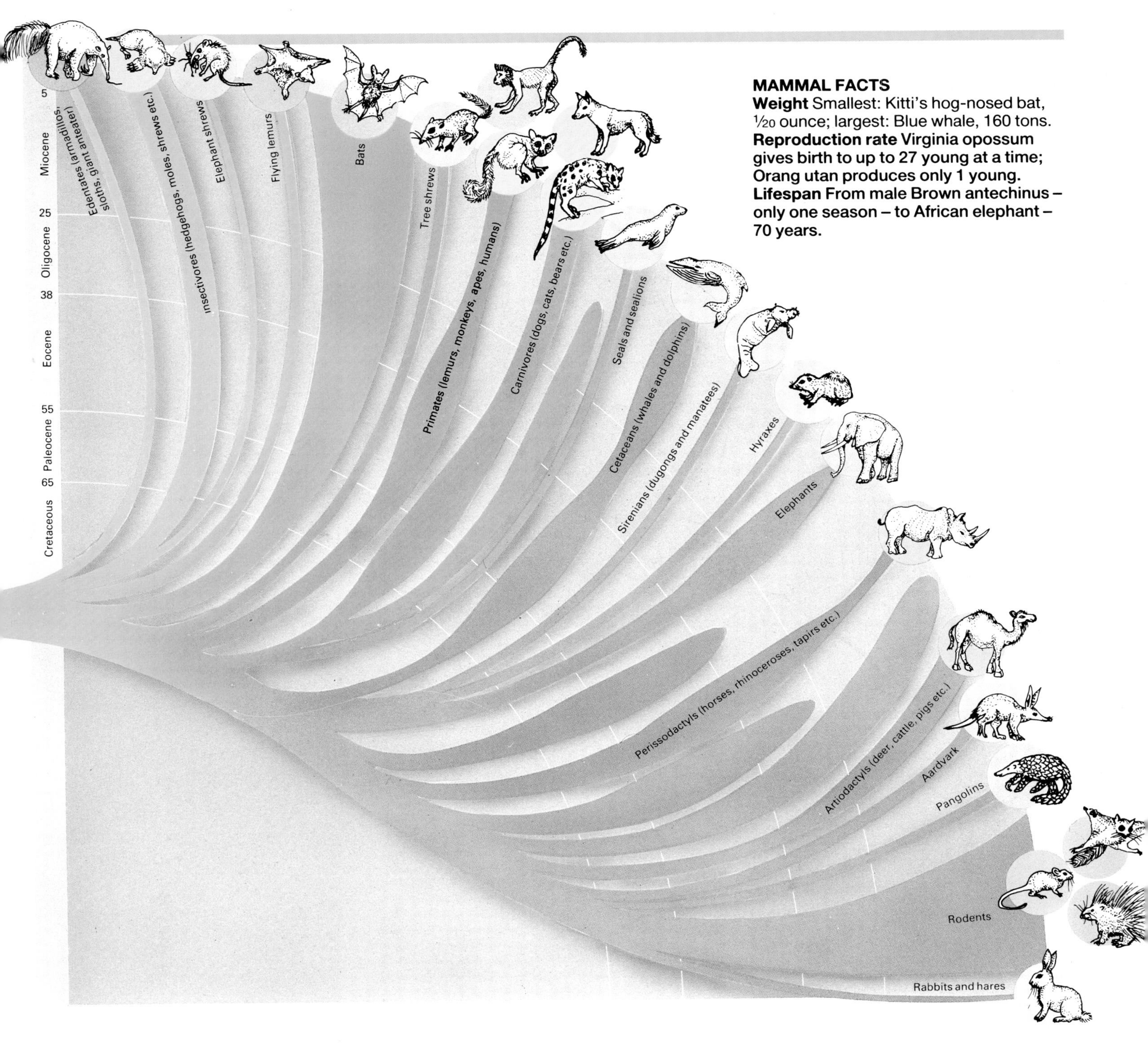

MAMMAL FACTS
Weight Smallest: Kitti's hog-nosed bat, 1/20 ounce; largest: Blue whale, 160 tons.
Reproduction rate **Virginia opossum gives birth to up to 27 young at a time; Orang utan produces only 1 young.**
Lifespan **From male Brown antechinus – only one season – to African elephant – 70 years.**

During these 300 million years, the continents have moved around the globe, and the climate has changed. Not all the mammals that arose were successful. Some branches of the tree do not reach to the present day. Others, like the rodents and bats, are still expanding and changing today.

for chasing after prey, seals have flippers for swimming, and bats have wings for flying.

A mammal's jaws and teeth reveal how it feeds. Lions have special sharp teeth for seizing prey, tearing flesh and crushing bones. Sheep have large flat teeth for grinding leaves. Baleen whales have huge plate-like sieves for filtering sea water.

Classifying the mammals according to their body structure naturally arranges them in groups with similar life styles and feeding habits.

BEAVERS

The crisp morning air echoes to the sound of chiselling as a beaver gnaws away at the trunk of a young tree. The tree grows thinner and thinner at the base until it snaps and crashes to the ground. Grasping the trunk in its powerful jaws, the beaver drags it to the water's edge and dives in. The tree is pushed and pulled into position in the sprawling mound of branches, twigs and mud which makes up the beaver's dam.

Beavers are rodents which have become adapted to living in water. Their thick fur is waterproofed with grease, while their hind feet are webbed for swimming. When beavers dive, their ears and nostrils close, and a membrane covers their eyes. The throat becomes blocked off by the tongue to prevent water filling the lungs.

Beavers swim using an up-and-down action of their broad flat tail to give extra force against the water. On land, they are much less agile. They have a clumsy, shuffling walk.

UNIQUE DAM-BUILDERS

The dam-building activity of beavers is unique in the animal world. Large branches or tree trunks form the framework of the dam, while the beavers pack sticks, stones and mud into place around them. This solid unit holds back river water so that it

▲This North American female beaver is carrying her kit through the water using her large teeth and front feet.

▼The Mountain beaver does not build dams. It lives in underground burrows in coniferous forests.

BEAVERS Castoridae and Aplodontidae (*3 species*)

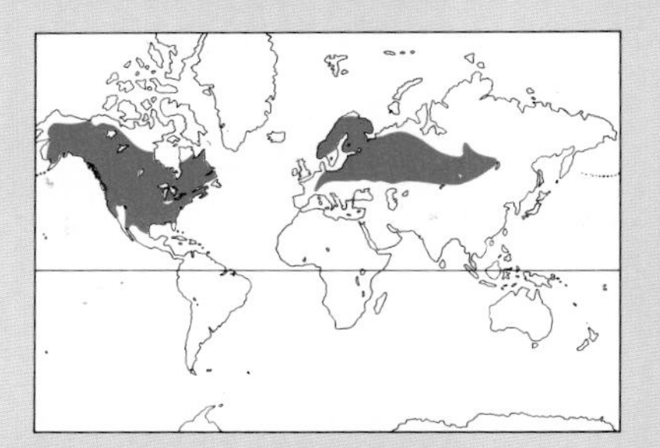

Habitat: swamp near lakes and streams.

Diet: leaves, herbs, grasses, woody stems.

Breeding: litters of 1–8 after pregnancy of 105 days.

Size: head-body 2½-4ft; weight 24-65lb.

Color: reddish-brown, sometimes darker.

Lifespan: 10-15 years.

Species mentioned in text:
European beaver (*Castor fiber*)
Mountain beaver (*Aplodontia rufa*)
North American or Canadian beaver (*Castor canadensis*)

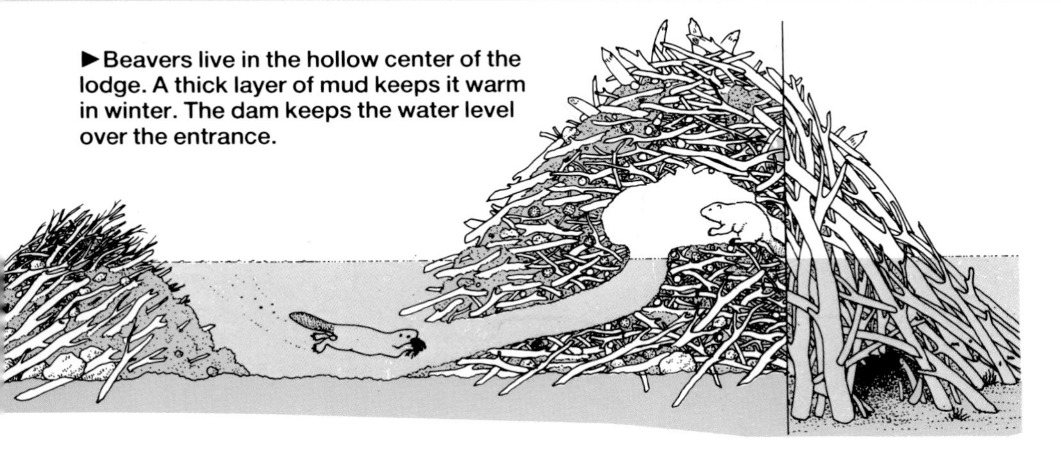

►Beavers live in the hollow center of the lodge. A thick layer of mud keeps it warm in winter. The dam keeps the water level over the entrance.

spreads out to form a deep pond.

Beavers are vegetarians, and in the fall they gather and store woody plant material in the pond. The cold temperature stops decay organisms attacking the material. The beavers rely on it for food throughout the winter. In spring and summer, when there is a greater choice of food, beavers change to a diet of soft leaves.

The beaver's living quarters or lodge is usually built in deep water behind the dam. It is also built of mud and branches. The entrance is under water to prevent enemies getting in, and it leads to a central chamber.

Male and female beavers stay together as a mating pair for several years, and both help to raise the young. They produce one litter each spring consisting of several kits. These are born covered in fur, able to swim and with their eyes open, but they do not leave the lodge until they are weaned at 6 weeks. The young beavers born the previous year help their parents to look after the kits, bringing fresh leaves to the lodge. They leave the family unit in their second year to start their own lodges.

TRAPPED IN THOUSANDS

Beaver skins are highly prized for making coats and hats because they are warm and waterproof. Thousands of beavers are trapped in Russia, America and Europe every year. They have almost been wiped out in the past by over-hunting. The European beaver is now found only in small isolated populations.

◄By creating deep ponds, beaver dams cause important changes in the local habitat.

▲Beavers use scent to mark their territory, so they have extremely sensitive noses.

SQUIRRELS

The Sun rises over the dry plains of central North America. The inhabitants of a small colony are already awake, preparing for the day ahead. Suddenly a male spots a puma and he barks out a warning. His companions dash back into their homes, while he stands erect on a look-out mound, keeping watch as he cleans his whiskers. This is, of course, not a human town. About 2,000 Black-tailed prairie dogs live here, in burrows under the short, dry grass.

All squirrel species grip small items of food firmly in their forepaws, like the Cape ground squirrel (above left). The European red squirrel leaves many feeding signs (above right) as it opens cones and nuts to extract the seeds.

It may seem odd to think of squirrels living underground rather than in trees. But the burrowing prairie dogs are just some of the members of the very numerous and widespread squirrel family. This family includes the familiar Gray squirrel and European red squirrel and other tree-dwelling species, the flying squirrels and the ground-dwelling prairie dogs, chipmunks and sousliks.

Squirrels, being so variable in their way of life and habits, are found in many different types of country – from mountain meadow to rocky cliff, dry grassland and tropical rain forest. Members of the family are found on every continent except Australia and Antarctica.

THE SQUIRREL'S MENU

A typical squirrel has a long rounded body, large eyes, strong legs and claws and a bushy tail. The long, chisel-shaped front teeth show that they are members of the rodent group. These teeth are well suited for chopping and opening nuts and seeds, snipping off flowers, fruits and shoots and levering strips of sappy bark from a tree trunk. Plant food is the main diet of most types of squirrel, as well as mushrooms and other fungi.

▶The European red squirrel is many people's idea of a typical squirrel, with its large eyes, pricked ears and long, bushy tail.

SQUIRRELS Sciuridae (*267 species*)

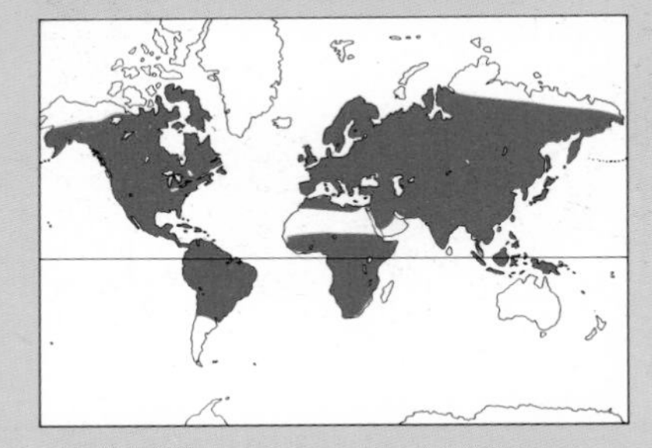

Habitat: tropical forest to woods, grassland and gardens.

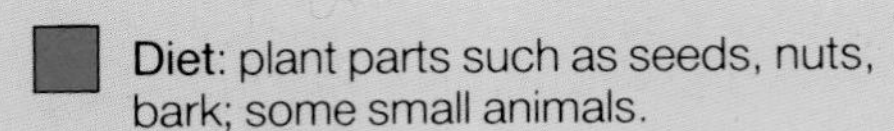

Diet: plant parts such as seeds, nuts, bark; some small animals.

Breeding: most species have one spring litter after pregnancy of 3-6 weeks; number of young varies: 1 or 2 (flying squirrels) to 9 or more (Gray squirrel).

Size: smallest (African pygmy squirrel): head-body 2½in, tail 2in, weight ⅓ ounce; largest (Alpine marmot): head-body 25in, tail 6½in, weight 17½lb.

Color: shades of red, brown or gray.

Lifespan: 1-2 years in smaller species, 8-10 years for some marmots.

Species mentioned in text:
African pygmy squirrel (*Myosciurus pumilio*)
Alpine marmot (*Marmota marmota*)
Belding's ground squirrel (*Spermophilus beldingi*)
Black-tailed prairie dog (*Cynomys ludovicianus*)
Cape ground squirrel (*Xerus inaurus*)
European red squirrel (*Sciurus vulgaris*)
Giant flying squirrel (*Petaurista* species)
Gray squirrel (*Sciurus carolinensis*)
Shrew-faced ground squirrel (*Rhinosciurus laticaudatus*)
Siberian chipmunk (*Tamias sibiricus*)
Woodchuck (*Marmota monax*)

One of the reasons for the squirrels' success is their broad diet. Many species eat beetles or other insects. Some are omnivorous, and their diet includes lizards, small birds, eggs and worms as well as plant food. The Shrew-faced ground squirrel is unusual in that it mostly eats insects, such as termites and caterpillars.

◀A female Belding's ground squirrel stands erect and calls to warn her neighbors of an approaching coyote.

LEAPING AND GLIDING

Tree squirrels are very agile animals. They can run straight up or down a tree trunk and leap several yards through the branches, clinging with their sharp claws. They have keen sight, which helps them judge distances accurately when scampering among the twigs.

Flying squirrels are specialized not for true flying but for gliding. There is a furred flap of skin down each side of the body, stretching between the front and back legs. As the flying squirrel jumps, it extends all four legs to stretch the skin and form a parachute. The tail is free to move from side to side, like a rudder for steering.

Flying squirrels, unlike other species, are active at night. When searching for food or avoiding a predator, they swoop from high in one tree to low on the trunk of another. The larger species, such as the Giant flying squirrel, can glide 330ft or more.

A WINTER'S TAIL

The squirrel's tail has many uses, depending on the species. In flying squirrels and tree squirrels it acts as a rudder when leaping, and as a counterweight when balancing on a branch. It may also be used as a signal, informing other squirrels of anger, acceptance or readiness to mate. Species living in hot places fluff out their tails over their backs, to act as a sun-shade. In winter, many tree squirrels sleep in their nests with their tails wrapped around them for warmth.

Gray, red and other tree squirrels do not hibernate. They may stay inside their stick-and-twig nests, called dreys, for several days during bad weather. But they emerge on fine winter days to feed and drink. However, many ground squirrels and marmots hibernate for about 6 months, when their body temperature falls to only a few degrees above freezing.

The woodchuck or groundhog, a marmot of North America, is a famous

▼Young ground squirrels (called "pups") leaving their burrow for the first time. These pups are about 27 days old.

▲Types of tree squirrel An African pygmy squirrel **(1)** carries a nut. An Abert or Tassel-eared squirrel (*Sciurus aberti*) **(2)** holds food in its forepaws. An Indian giant squirrel (*Ratufa indica*) **(3)** leaps to a nearby branch. An American red squirrel (*Tamiasciurus hudsonicus*) **(4)** hangs by its hind claws while cracking a nut. A Southern flying squirrel (*Glaucomys volans*) **(5)** glides from its tree hole. Prevost's squirrel (*Callosciurus prevosti*) **(6)** keeps watch from a sawn-off branch.

hibernator. Legend says that it wakes on February 2nd each year – Groundhog Day. If it sees its shadow, this means sunny cold weather, and it returns to its burrow for another 6 weeks. (This story is unfounded.) When it emerges in spring, it is very thin and has lost up to half its weight.

MAKING A LARDER
Squirrels are famous storers of nuts, seeds and other food to last them through the winter. The Siberian chipmunk may store up to 13lb of food, perhaps 100 times its body weight, in its burrow. This is its source of food when the chipmunk wakes in spring, before new plant growth becomes abundant.

Tree squirrels bury food or hide it in undergrowth, to dig up in winter. European red squirrels can smell pine cones they have buried 1ft below the surface. Gray squirrels may carry acorns 100ft from an oak tree before hiding them in a hole. But many squirrels forget where they have buried food, so their activities help to spread seeds and to plant them for future growth.

BREEDING AND SOCIAL LIFE
Most squirrels, like other rodents, are able to breed within a year of being born. There is usually one litter per year. Newborn young are blind and helpless but develop fast, and within 2 months they are ready to leave their nest and fend for themselves.

Most tree squirrels live alone. Each has a territory and chases away intruders. The territory of a male may overlap those of several females. During spring, boundaries break down, and animals come together for a few days to mate.

Ground squirrels tend to be more social and live in groups. Fifty or more Alpine marmots may occupy a large burrow system, with males, females

▲Types of ground squirrel A Siberian chipmunk with cheek pouches full of food **(1)**. An Alpine marmot stands upright and gives an alarm whistle as it spots a wolf **(2)**. A Shrew-faced ground squirrel, also called the Long-nosed squirrel, extracts termites from a hole in a fallen branch **(3)**. A Western ground squirrel (*Xerus erythropus*) indicates worry by arching and fluffing its tail **(4)**.

◀The female Belding's ground squirrel carries up to 50 loads of dry grass to line her nest before giving birth.

▶Black-tailed prairie dogs relax in the Sun. The mound around their tunnel entrance helps to prevent the tunnel flooding.

and young living as one large family. Prairie dogs form even bigger colonies, called "townships," of perhaps 5,000 animals. Each township consists of several family burrows called "coteries." In summer, prairie dogs from different coteries are friendly to each other. In winter, when food is scarcer, coterie members defend their burrows and territories.

SQUIRREL DAMAGE

Two species of squirrel are officially listed as being endangered. But squirrels in general are common, and some species are regarded as pests in certain areas. Prairie dogs used to cause great damage to crops and destroy grassland grazed by farm animals. Poisoning has now made some of them scarce except in remote areas and national parks. The Gray squirrel, in particular, is unloved by foresters, because it strips bark from young trees, probably to eat the sweet-tasting sapwood – but also killing the tree in the process. Guns and traps may reduce its numbers for a time, but there seems to be no permanent answer.

GOPHERS

Summer drought has hit the south-western USA. In the dry, crumbly soil, a gopher digs for roots and succulent tubers. But there is little food left. It decides to chance a raid on a neighboring gopher's territory, where a juicy cactus grows. The neighbor is waiting, and it rears up and bares its teeth and claws to defend its food source.

The gopher is often called the pocket gopher because it has "pockets" on its cheeks. These are loose flaps of furry skin, folded forwards and down on each side to form pouch-like containers. When foraging, the gopher pushes food into its pockets using its forelegs. It carries the food back to its burrow, to store and eat later.

The gopher's skin is loose and floppy, its fur is short and thick, and its ears are small. These features suit it to life underground, running and turning in long, narrow tunnels. Gophers can run backwards almost as fast as they run forwards.

PICK AND SHOVEL

Usually a gopher digs with its strong forelegs, using its claws like shovels. But in hard soil it may also use its teeth like a pickaxe, chiselling along and eating any roots, bulbs and other plant food it comes across. Both teeth and claws grow up to 1/25in a day, to keep up with tunnelling wear and tear.

Most gopher species live in North and Central America. Some are called taltuzas, and there are many local names, such as digger rats. Although all species look generally similar, there can be great variety even within one species. In some areas the Valley gopher is pale fawn, while in other places it is almost black. In most species, males are larger than females.

Each gopher has its own system of tunnels in its home territory. Normally, individuals defend their burrows and territories – except in the breeding season, usually in spring. At

▼This Valley gopher shows the large claws on its front feet which make good, spade-like digging tools.

GOPHERS Geomyidae
(*34 species*)

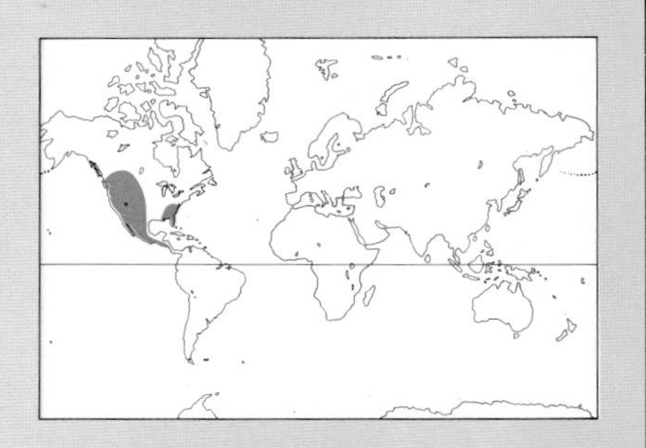

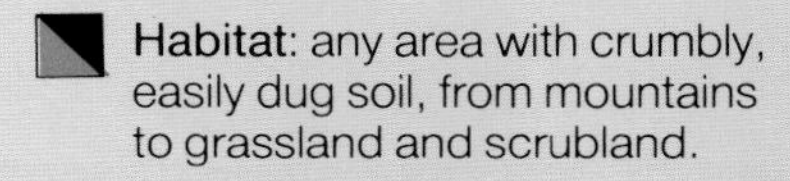

Habitat: any area with crumbly, easily dug soil, from mountains to grassland and scrubland.

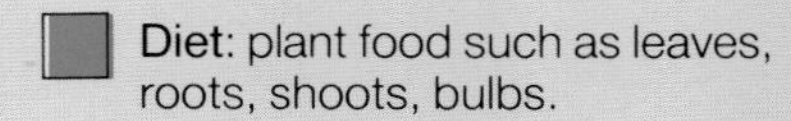

Diet: plant food such as leaves, roots, shoots, bulbs.

Breeding: varies, average litters of 2-5 after pregnancy of 17-20 days.

Size: western gophers: from head-body 5in, weight 1½ ounces; Large pocket gopher: up to head-body 12in, weight 2lb.

Color: varies between and within species, from pale fawn to black.

Lifespan: 1-4 years.

Species mentioned in text:
Large pocket gopher or taltuza (*Orthogeomys grandis*)
Valley or Valley pocket gopher (*Thomomys bottae*)

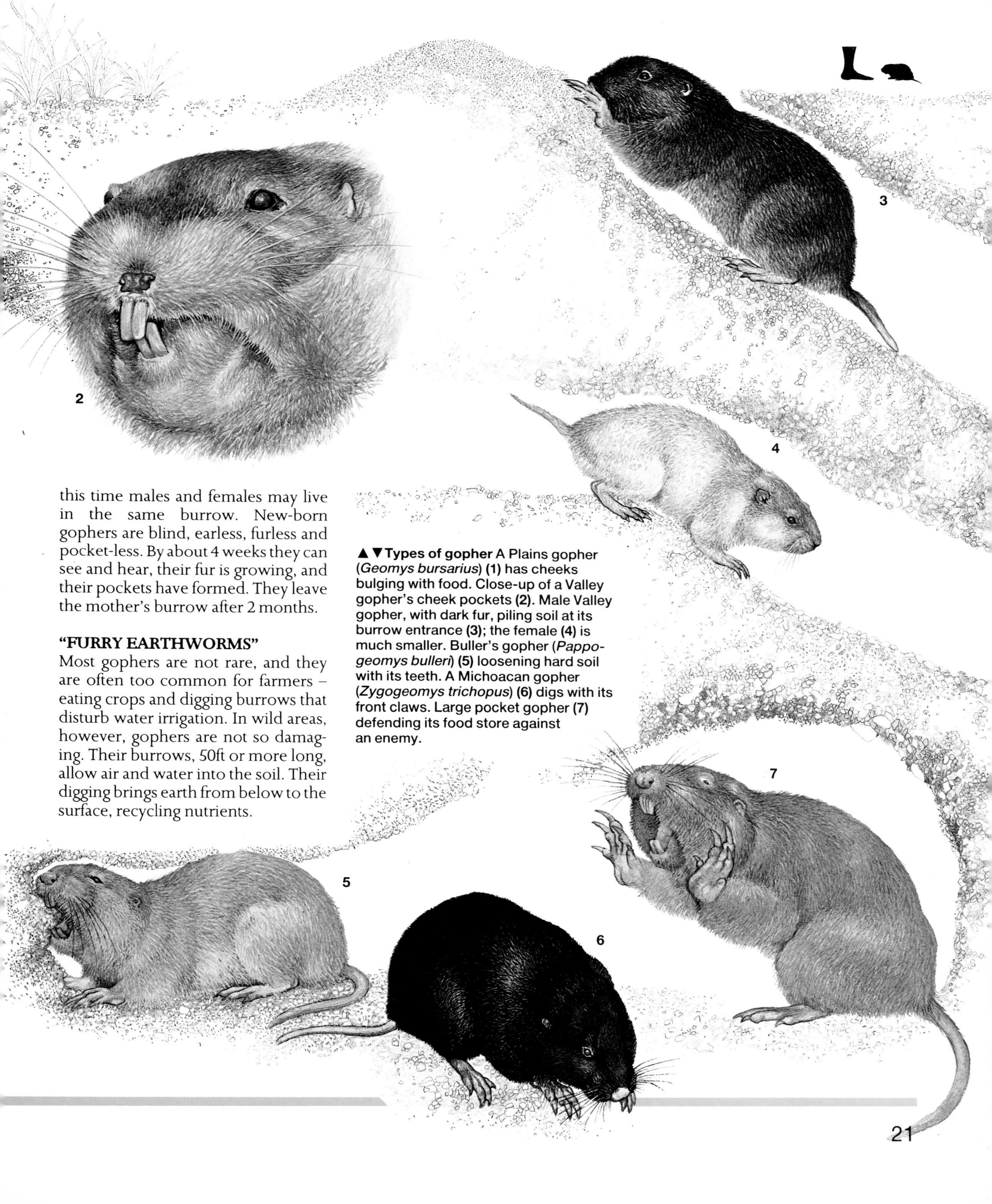

this time males and females may live in the same burrow. New-born gophers are blind, earless, furless and pocket-less. By about 4 weeks they can see and hear, their fur is growing, and their pockets have formed. They leave the mother's burrow after 2 months.

"FURRY EARTHWORMS"

Most gophers are not rare, and they are often too common for farmers – eating crops and digging burrows that disturb water irrigation. In wild areas, however, gophers are not so damaging. Their burrows, 50ft or more long, allow air and water into the soil. Their digging brings earth from below to the surface, recycling nutrients.

▲▼**Types of gopher** A Plains gopher (*Geomys bursarius*) **(1)** has cheeks bulging with food. Close-up of a Valley gopher's cheek pockets **(2)**. Male Valley gopher, with dark fur, piling soil at its burrow entrance **(3)**; the female **(4)** is much smaller. Buller's gopher (*Pappogeomys bulleri*) **(5)** loosening hard soil with its teeth. A Michoacan gopher (*Zygogeomys trichopus*) **(6)** digs with its front claws. Large pocket gopher **(7)** defending its food store against an enemy.

SPRINGHARE

In the heat of the African noon, a female springhare rests in her cool underground burrow. A young springhare, a miniature version of its mother, suckles quietly. Above ground, a mongoose approaches, sniffing the air. It locates the burrow entrance and silently slips inside in search of the young animal. The female springhare is instantly alert. She lashes out at the mongoose with her powerful hind feet and attacks it with her large front teeth. The mongoose retreats.

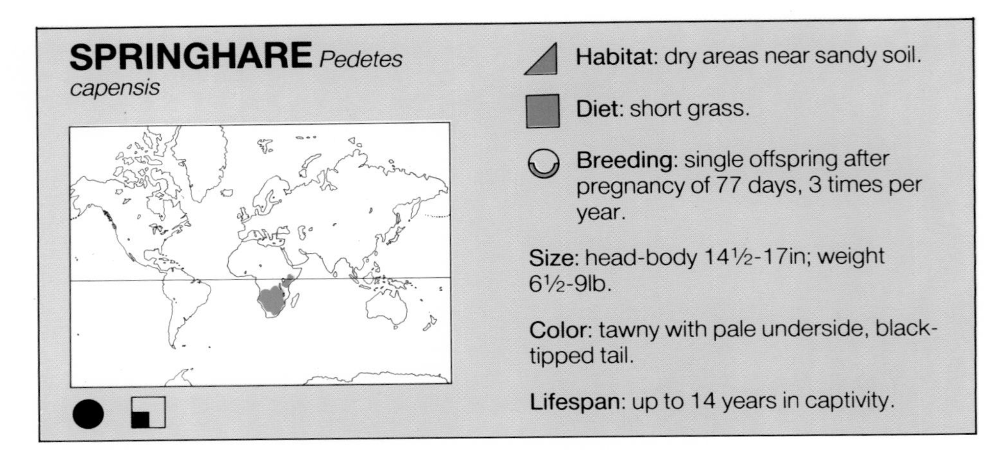

SPRINGHARE *Pedetes capensis*
Habitat: dry areas near sandy soil.
Diet: short grass.
Breeding: single offspring after pregnancy of 77 days, 3 times per year.
Size: head-body 14½-17in; weight 6½-9lb.
Color: tawny with pale underside, black-tipped tail.
Lifespan: up to 14 years in captivity.

►A successful hunt for these Bushmen means death for a springhare. No part of its body will be wasted.

▼Springhare behavior: standing **(1)**, leaping forward **(2)**, feeding **(3)** and grooming **(4)**.

Springhares are not true hares, even though they have a rabbit-like head. They are, in fact, rodents related to the squirrel and beaver, with well-developed gnawing teeth. They hop along quite rapidly on their long hind feet like small kangaroos.

If chased by an enemy, springhares can leap an amazing 10ft or more. The brush-tipped tail, which is longer than the body, is used to give them extra balance. The claws on the hind feet

are very tough and hoof-like. The front feet are much smaller, with more delicate claws for grooming the longish fur.

GRASSLAND FEEDERS

Springhares have a similar diet to rabbits. Their main source of food is grass, and they can survive on the shorter grasses which the larger hoofed animals cannot cope with. They select only the youngest, succulent blades which are rich in protein.

Unfortunately for farmers, springhares cannot tell the difference between grass and the green tips of newly planted crops like sweet corn. They can cause serious damage, and farmers destroy them as pests. Native Bushmen also trap them for food, because they are a useful source of protein. The skins are used for making bags and clothing.

Unlike rabbits, female springhares become pregnant only about three times a year and give birth to single offspring. This allows each female to devote all her energy to feeding and protecting the one baby. Animals which produce litters of several young tend to lose most of them to predators and starvation.

SAFETY OF THE BURROW

Young springhares are born covered in fur and with their eyes fully open, so they are active and on the move almost immediately. Each one lives alone in a burrow with its mother, and they do not venture above ground to feed until they are partly grown. Below ground they are safe from hunting birds like owls, as well as the big cats, but they still fall prey to snakes and mongooses.

Every burrow has more than one entrance so that the springhare can escape if one passage is blocked by an enemy. Some entrances are deliberately blocked with soil, but this can be quickly dug out if necessary.

The size of springhare populations is, despite human and other enemies, mainly controlled by the availability of food. In poorly grassed areas there are a much smaller number than in places where grass is more plentiful. They usually forage for food at night, in small groups of four to six, because this gives them greater protection against attack.

◀Prominent eyes and sensitive mobile ears allow the springhare to scan its surroundings for enemies.

RATS

On a Pacific island a Galapagos rice rat clambers through a bush in search of food. Its long whiskers quiver, and its keen eyes and ears strain to detect danger. Hearing the movements of another animal, lower in the bushes, the rice rat cautiously investigates. It is another rat, but bigger, browner and bolder – a Norway rat. Out of nowhere, a cat pounces. The Norway rat, fast and fierce, gets away. The Galapagos rice rat is less lucky.

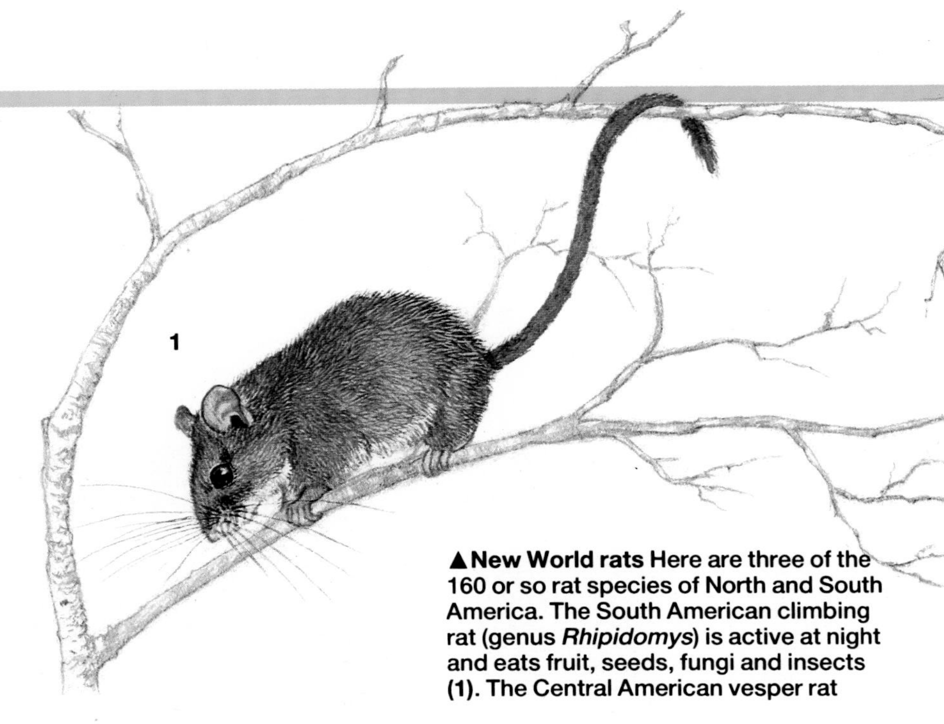

▲New World rats Here are three of the 160 or so rat species of North and South America. The South American climbing rat (genus *Rhipidomys*) is active at night and eats fruit, seeds, fungi and insects (1). The Central American vesper rat

What is the difference between a rat and a mouse? It is mainly size. Most rats – but not all – are larger than most mice. There are no important biological differences. Whether a species is called a rat or a mouse depends partly on local names and traditions. Rats and mice all belong to the same rodent family, Muridae, along with voles, lemmings, hamsters and gerbils. Altogether this family contains 1,082 species, in 241 genera, which is one-quarter of all mammals.

FOUND WORLD-WIDE

Rats live all over the world except in such cold climates as exist in the far north and on especially high mountains. A few species are very familiar and widespread since they live alongside people, in or near buildings, rubbish heaps and anywhere else there might be shelter and scraps of food. These include the Norway rat (also called the Brown, Common or Sewer rat) and the Roof rat (also called the Black or Ship rat).

Hundreds of other species of rat live away from people, in rocky mountains, deserts, woodland, grassland and tropical forest. A few are so rare that they are known only from museum specimens or one or two individuals caught in the wild. At least two species of rice rat, from the Caribbean islands, have become extinct. Many other rice rats are threatened – by other rat species brought by people, as well as by cats, dogs and other animals.

RATS Muridae (*450-500 species*)

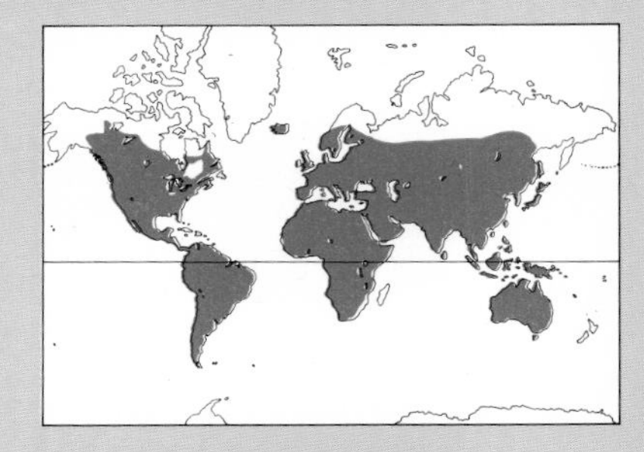

● ▪ ☠

Habitat: any land habitat except high mountains, including deserts, steppe, woodland and tropical forest.

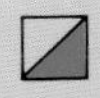

Diet: seeds, shoots, leaves and other plant food, also insects and other small animals.

Breeding: litters of 3-7 after pregnancy of 20-50 days; in suitable conditions litters may be every 3-4 weeks.

Size: smallest: mouse-sized (see page 30); largest (Cuming's slender-tailed cloud rat): head-body 19in, tail 13in, weight 4½lb.

Color: varies from cream or pale fawn through red, brown and gray to black; ears, nose, feet and tail usually pink-gray; some species have striped backs.

Lifespan: from only a few months up to 2-3 years in bigger species.

Species mentioned in text:
African swamp rats (genus *Otomys*)
Australian water rat (*Hydromys chrysogaster*)
Cuming's slender-tailed cloud rat (*Phloeomys cumingi*)
Galapagos rice rat (genus *Nesoryzomys*)
Giant blind mole-rat (genus *Spalax*)
Lesser bandicoot rat (*Bandicota bengalensis*)
Muller's rat (*Rattus muelleri*)
Multimammate rat (*Praomys natalensis*)
North American woodrat (*Neotoma micropus*)
Norway rat (*Rattus norvegicus*)
Polynesian rat (*R. exulans*)
Roof rat (*R. rattus*)
Zokor (genus *Myospalax*)

(*Nyctomys sumichrasti*), also nocturnal, eats fruit and is completely at home in trees, even nesting there (2). The Central American climbing rat (genus *Tylomys*), another nocturnal rat, prefers trees along river banks and lake shores (3).

THE TYPICAL RAT

The Norway rat is a typical rat. It has a body up to 10in long, coarse browny-gray fur, a long snout, beady eyes, smallish pinky ears with fine hairs, and strong legs with stout claws. Its tail is furless, up to 8in long.

Rats, like mice, have four sharp front teeth called incisors, two in the upper jaw and two in the lower. They also have 12 crushing cheek teeth, three on each side of the upper jaw and three on each side of the lower. While this rodent-like design evolved originally for eating seeds and other plant food, many rats eat whatever they find. The Norway rat takes almost anything, plant or animal, living, dying, dead or rotting. Such a varied diet is one reason why this species is so widespread and numerous.

Another reason for the Norway rat's success is its breeding rate. As with many other species of rat (and mouse), when food and shelter are plentiful a female can have several litters each year. She lives in a burrow and makes a nursery nest of grass, straw, bits of wood and paper, rags and other material. On average, the Norway rat has about 7 young in each litter, but some litters number 10 or more. So one female can raise more than 50 young in a "good" year. However, as many as 50 per cent of these will die from predation or lack of food before they reach breeding age (about 12 weeks). And 95 per cent of adults do not live for more than a year.

The Norway rat came originally from the Caspian Sea area. It began to spread west in the 11th century, and by the early 18th century had reached Britain. It swims well. In the water it is often mistaken for a water vole.

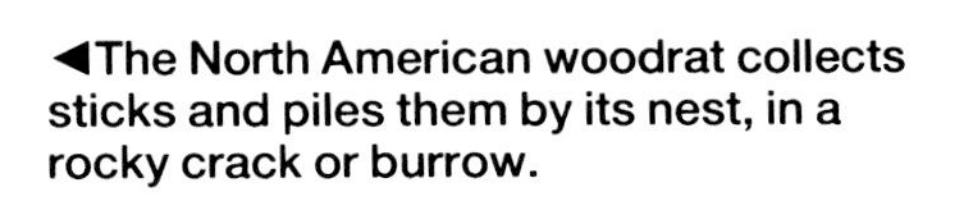

◀The North American woodrat collects sticks and piles them by its nest, in a rocky crack or burrow.

ANOTHER SEAFARING RAT
Another international rodent is the Roof or Black rat. It is not always black – some individuals are brown, gray or even dirty cream. It has a smaller body, larger ears and a longer tail (greater than its body length) than the Norway rat.

The Roof rat came originally from India. Like the Norway rat, it travelled on ships and in cargoes, reaching western Europe in the 11th century. Being a tropical creature by nature, it tended to stay near warm and sheltered places. When the bigger, bolder Norway rat arrived, the Black rat became less common.

Today, both species live in many countries across the world. The Roof rat tends to be more common in tropical regions, infesting towns and villages. The Norway rat is more numerous in temperate areas and in cities and ports in the tropics.

RATS OF TWO WORLDS
The Norway and Roof rats belong to the group known as Old World rats and mice (Murinae). There are 406 species, of which about 200-250 are regarded as rats. They inhabit all of Europe, Africa, central and southern Asia and Australia. Some species live mainly in burrows, others spend almost all their lives in trees. All are chiefly herbivores, eating seeds, shoots, fruits and leaves.

There are more rat species in the warmer and damper regions of Africa and Asia than there are in Europe or in dry or cold areas. There are also many species in the islands of South-east Asia. In the Philippines alone there are 30 rat species found nowhere else. They include Cuming's slender-tailed cloud rat (the largest rat species) and two other large species. In New Guinea there are another six sizeable species.

For the past 30 million years, the New World rats and mice (Hesperomyinae) have evolved separately from their Old World cousins. There are now 366 species, of which about 160 are commonly called rats. Like their counterparts in the Old World, they live a variety of life-styles.

The Old and New World rats also show a similar range of adaptations. In climbing species the tail has become a long balancing or a grasping organ. The big toes of hands and feet of tree-living species are often opposable, allowing the animals to grasp thin branches.

OTHER RAT GROUPS
Related to the Old and New World rats are about eight other groups of rats. The African swamp or vlei rats live in central and southern Africa, especially in upland areas. They have blunt noses, long fur and short tails, and look more like voles than typical rats. They prefer damp places and eat grass and other tough-stemmed plants.

In South-east Asia and Australia there are about 20 species of Australian water rats that are well suited to life in marshes and rivers. The largest is the Australian water rat itself, which may weigh more than 2½lb. It has webbed hind feet and hunts underwater by day for fish, frogs, crabs, insects and shellfish. It is often mistaken for a platypus.

The zokors are burrowing, vole-like rats found in dry grassland and woods in central Asia and China. They have small eyes and ears and very large front claws for digging. They eat roots and bulbs, sometimes surfacing to forage for seeds. Zokors collect huge underground stores of food for the winter.

UNDERGROUND RATS
Perhaps the oddest rats are the blind mole-rats of Africa, the Middle East and Western Asia. The largest is the

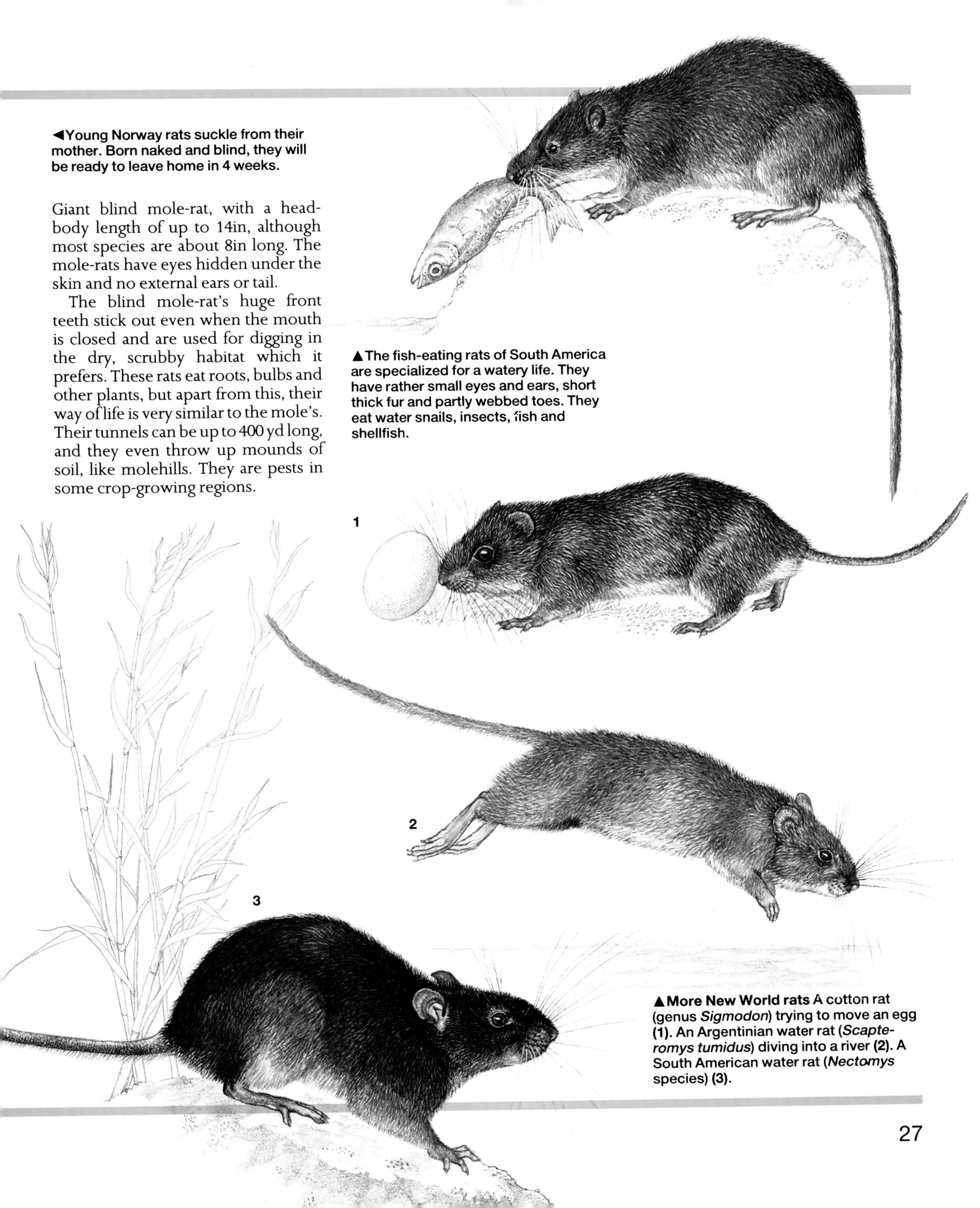

◀Young Norway rats suckle from their mother. Born naked and blind, they will be ready to leave home in 4 weeks.

Giant blind mole-rat, with a head-body length of up to 14in, although most species are about 8in long. The mole-rats have eyes hidden under the skin and no external ears or tail.

The blind mole-rat's huge front teeth stick out even when the mouth is closed and are used for digging in the dry, scrubby habitat which it prefers. These rats eat roots, bulbs and other plants, but apart from this, their way of life is very similar to the mole's. Their tunnels can be up to 400 yd long, and they even throw up mounds of soil, like molehills. They are pests in some crop-growing regions.

▲The fish-eating rats of South America are specialized for a watery life. They have rather small eyes and ears, short thick fur and partly webbed toes. They eat water snails, insects, fish and shellfish.

▲**More New World rats** A cotton rat (genus *Sigmodon*) trying to move an egg **(1)**. An Argentinian water rat (*Scapteromys tumidus*) diving into a river **(2)**. A South American water rat (*Nectomys* species) **(3)**.

PESTS AND DISEASES

Most species of rat are rarely seen by people, but a few are pests. Besides the Norway rat and Roof rat, they include the Multimammate rat in Africa, the Polynesian rat in Asia and the Lesser bandicoot rat in India.

All these rats, with their sharp, ever-growing incisors, chew away water-pipes and drains, making them leak, and gnash through electricity cables, causing power cuts. They gnaw their way into food stores (especially grain), eating what they can and contaminating the food with their urine and droppings. Their burrowing undermines walls and foundations – buildings have collapsed due to warrens of rat tunnels beneath.

Rats also carry disease. In the 14th century millions of Europeans died from bubonic plague, the Black Death. The bacteria (germs) causing the plague were carried by Roof rats. Fleas sucked blood from the rats, took in the bacteria and then bit people, passing on the disease. A number of serious diseases, including lassa fever, rat-bite fever, Weil's disease (leptospirosis) and a form of typhus, are today spread by the bite, urine, droppings and parasites of rats.

Pest rat species have been hunted, poisoned and trapped over the centuries. Millions have been killed. Yet they are adaptable and agile, able to eat almost anything and breed very quickly. Many species have now acquired resistance to conventional rodent poisons, such as warfarin. In the sewers and underground tunnels of most large cities, the rat population continues to increase. It is unlikely we will ever be able to banish them completely from areas where people live and work.

►A Muller's rat of warm damp regions of the Far East shows the typical rat's agility as it grasps with its sharp claws and balances using its tail.

MICE

On a wheat farm in southern Canada laborers arriving for work one morning find that most of the season's harvest has been destroyed overnight by mice. The mice have chewed and broken their way into the sacks of grain, and either eaten the contents or contaminated it with their urine and droppings. The farmer calls in the pest control department, which taints the rest of the grain with poison. Within two days, several thousand mice are killed. The only problem that remains is to remove the poisoned mice before other animals eat them.

A mouse is a small rat – or, to put it the other way round, a rat is a large mouse. There is no real biological difference between mice and rats. Scientists include them all in the same family (Muridae). Of the 1,082 species in this family, between 350 and 400 are commonly called mice. They live all over the world except on high mountains, and in lakes and oceans.

Most experts agree that the House mouse, which has followed people to every continent, is the world's most numerous mammal. The Australian farmer who found 28,000 House mice dead on his veranda, after putting down poison overnight, would probably agree.

BUNDLE OF ENERGY

The Wood mouse of Europe and Asia is a typical mouse. It has a head and body about 3½in long, and a finely haired, scaly-looking tail of about the same length. Its big, black, beady eyes and large fine-haired ears indicate that it is active by night. It also has a keen sense of smell and many large, sensitive whiskers on its snout. The fur is generally chestnut or sandy-brown on the back and much lighter, even white, on the underparts, with a small yellow streak on the chest.

The Wood mouse is a very active animal, scrabbling and bounding about as it forages over its territory. Like most mice, it lives alone. It sniffs new objects cautiously and is always looking and listening for predators. Even so, Wood mice are commonly caught by foxes, cats and owls.

▲The American harvest mouse, like its Old World cousin, builds ball-shaped grass nests near the ground.

◀The mole mice of South America live in underground burrows and eat mainly termites and other insects.

MICE Muridae (*400 species*)

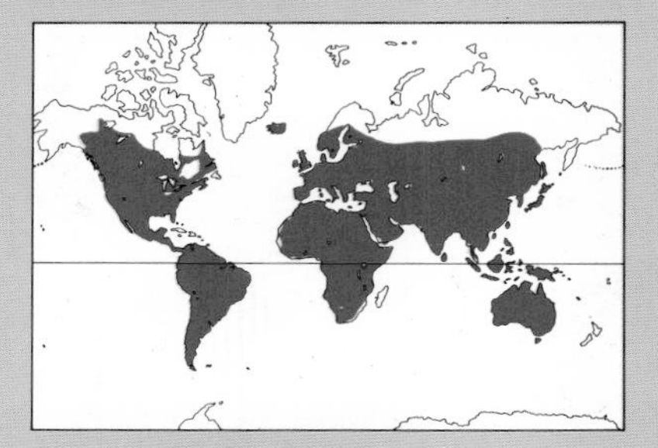

Habitat: any land habitat except high mountains, including deserts, steppe, woodland and tropical forest.

Diet: seeds and other plant food, also some animal food; a few species completely insectivorous.

Breeding: usually 1-3 litters per year; litters of 2-5 after pregnancy of 3-4 weeks; in suitable conditions litters monthly in some species.

Size: smallest (pygmy mice): head-body 2in, tail 1in, weight ¼ ounce; largest the size of a small rat (page 24).

Color: cream or pale fawn through red, brown and gray to almost black, often with paler underparts; tail may be naked or furry; some species have striped backs.

Lifespan: from a few months to 2 years.

Species mentioned in text:
American harvest mouse (*Reithrodontomys humilis*)
Climbing wood mouse (*Praomys alleni*)
Deer mice (*Peromyscus* species)
Fawn hopping mouse (*Notomys cervinus*)
Grasshopper mice (*Onychomys* species)
Harvest mouse (*Micromys minutus*)
House mouse (*Mus musculus*)
Larger pygmy mouse (*M. triton*)
Leaf-eared mice (*Phyllotis* species)
Mole mice (*Notiomys* species)
New World pygmy mice (*Baiomys* species)
Peter's striped mouse (*Hybomys univittatus*)
Pygmy mouse (*Mus minutoides*)
Shrew mouse (*Blarinomys breviceps*)
Volcano mouse (*Neotomodon alstoni*)
Water mice (*Rheomys* species)
Wood or Long-tailed field mouse (*Apodemus sylvaticus*)
Yellow-necked field mouse (*A. flavicollis*)

BREEDING LIKE MICE

Despite its name, the Wood mouse is found in a variety of habitats, including woodland, scrubland, hedge, moor, mountain and sand dunes, and in gardens and around outbuildings. It is often mistaken for the House mouse, although it does not have the strong smell of the House mouse.

The Wood mouse digs a burrow and hides in it by day, emerging to feed at night. It feeds mainly on buds and seeds, including nuts, which it is expert at opening with its incisor teeth. It stores food in the burrow, and the female gives birth there, in a nest lined with grass and moss. She may have four or five litters each year through the spring and summer, with about five young in each. So numbers can build up quickly, and this species is sometimes a pest around farms and where food is stored.

►The Harvest mouse's nest, about 3in across, is woven usually on grass stems or sometimes into a thorn bush.

▼A Yellow-necked field mouse scurries past rose hips by moonlight, carrying a hazel nut in its sharp front teeth.

LIFE IN TALL GRASS

Another European species is the Harvest mouse. This is one of the smallest rodents, with a body weight of only 1/4 ounce when fully grown. It lives in fields of crops and tall grass along hedges, highways and railway and canal embankments.

The Harvest mouse is famous for its tennis-ball-sized nests, which it weaves from living, shredded grass leaves and stems. It rests in such a nest, or in a burrow or hedgerow, between bouts of great activity – 3 hours sleeping and 3 hours feeding. This means the Harvest mouse is active by day and so it often falls prey to hawks, weasels, snakes and other hunters.

Young Harvest mice are born in a special nursery nest, more strongly built than the usual resting nest, between 4in and 12in from the ground. After a pregnancy of 21 days the female gives birth to between 5 and 8 young (rarely 12 or more). Their eyes open at 8 days, they begin to leave the nest after 12 days, and they are frequently independent of their mother at 15 days.

OLD WORLD MICE

The mice species mentioned so far are only some of the 150 to 200 species of Old World mice that live in Europe,

▲Newborn mice are hairless and cannot see or hear. But within 3 weeks they will be able to fend for themselves.

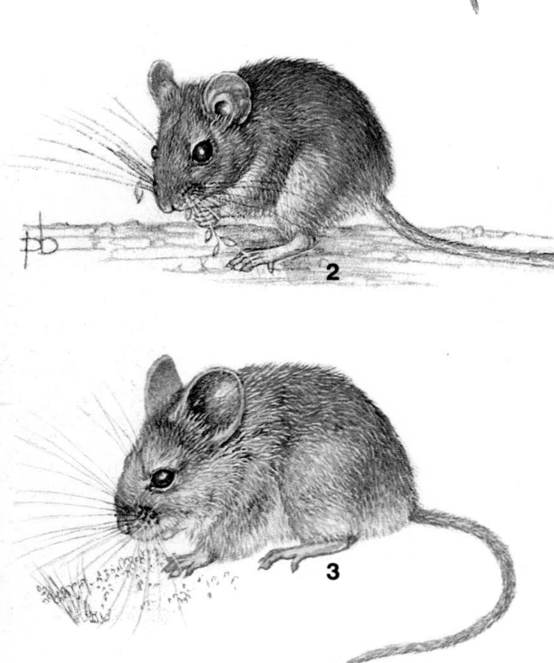

►**New World mice** A Pygmy mouse from southern North America **(1)**; Deer mouse from central North America **(2)**; Leaf-eared mouse from central South America **(3)**.

▼Leaf-eared mice, unlike most mice, are active by day. They often bask on sunny rocks, listening intently.

Africa, Asia and Australia. They belong to the Murinae group.

Many mice live in tropical forests. One research project in Uganda found four species of mice (as well as seven rat species) living together in one small area. The Pygmy mouse stayed on the ground and ate a variety of plants and small animals. So did the Larger pygmy mouse, although it tended to take larger food items. The Climbing wood mouse lived among the branches of low bushes, seldom coming to the ground or going higher than about 24in. Peter's striped mouse lived in taller branches and ate plant food. Living at different levels and taking different types of food, these mice were able to exist together.

MICE OF THE AMERICAS

About 200 mice species belong to the New World mice and rats (Hesperomyinae). Among the most familiar are the deer or white-footed mice of North and Central America. The various species of deer mice live on the ground and in trees, in all types of habitat from deserts to rocky hills and woodland. Their way of life is similar to that of the Old World mice.

The Volcano mouse is a close relative of the deer mice. It lives on the steep volcanic slopes of central Mexico, at heights of up to 14,000ft, and hides in burrows and crevices in and among the rocks.

Some of the most unusual mice are the water mice of Central America. They live in mountain streams and feed on water snails and fish. Large-bodied for mice, with a head-body length of up to 8in, they have webbed hind feet with bristly hairs on the outside, to aid swimming.

►Grasshopper mice, which range from southern Canada to central Mexico, give a shrill squeak when they see one another. This may help to keep them evenly spaced.

Also in South America live the burrowing and mole mice. These have large front claws, up to ⅓in long, for efficient digging. One species, the Shrew mouse, has tiny eyes, and its ears are hidden in its thick fur. It spends its life burrowing in the forest floor, eating mainly insects and other small creatures.

CLIMBING AND BURROWING

Apart from the two major groups, Old World and New World mice, there are also other, smaller groups of mice. One of these is the African climbing mice (Dendromurinae), which live mainly in grassland areas of Central and Southern Africa. New species of these animals are discovered every few years, and among them are some very distinctive mice.

The climbing mice themselves are agile, with long tails and slender feet. Their way of life is much like that of the Harvest mouse, and some species build ball-shaped nests of grass. They feed on seeds, fruits and small animals such as beetles, lizards and birds.

Also in this group are the fat mice, which make large burrows. Before the dry season they feed eagerly and put on thick layers of body fat. During the dry season they stay in their burrows, in a form of hibernation. They also do this during the day in the wet season. Mice are generally such busy creatures, but the fat mice are inactive in the extreme.

OF MICE AND MEN

Most species of mice have a high breeding rate. It is one of their survival methods, since they are small and relatively defenseless animals. Another means of survival is their wide choice of food. For example, Harvest mice eat seeds, fruits, buds, shoots and some insects such as small beetles. Wood mice also have a varied diet, feeding on seeds and other plant food, worms, grubs and snails – after nibbling through the shells.

House mice are even less fussy and make meals of lard, butter, soap, candle wax, frozen foods such as meat and vegetables, paper, cardboard, leather and much else besides. This is one reason why they are so numerous and widespread. Other reasons are that in many parts of the world they have become resistant to normal rodent poisons and, if disturbed, they will readily leave one home and establish another without stress.

Besides being pests, mice have also given benefit to people. They have long been kept as pets, and more than 3,000 years ago "mouse worship" was practised in Western Asia. Today strains of mice with white, black and patched fur, bred originally from the House mouse, are used in laboratory experiments for testing new medicines and chemicals. This small rodent is one of the most useful, as well as one of the most destructive, creatures on Earth.

▼The 49 species of deer mice are widespread across North America. They eat seeds and are active at night.

►The House mouse has lived alongside people for thousands of years, perhaps since the first farmers of the Middle East cultivated and stored grain.

▼**Old World mice** A Harsh-furred mouse (*Lophuromys sikapusi*) from Africa eats an insect **(1)**. Spiny mice (*Acomys* species) have strong, spiky hairs on their backs **(2)**. The Pencil-tailed tree mouse (*Chiropodomys gliroides*) has broad feet for a good grip and a long, tufted tail for balance **(3)**. The Fawn hopping mouse of Australia has whiskers that measure more than half its body length **(4)**. A Four-striped grass mouse (*Rhabdomys pumilio*) shows its camouflaging stripes while cleaning **(5)**.

VOLES AND LEMMINGS

In late summer a Norway lemming noses its way down a bank. It is leaving to find new pasture. The grasslands here have become too crowded with others of its kind. More and more lemmings join the procession, all migrating from over-populated homelands. As they march they are funnelled into a strip of land sticking into a lake. Lemmings can swim, and they plunge in, hoping for fresh grass on the far side. But the lake is too wide, and they all drown.

▼A Bank vole nibbles an acorn held by its forefeet. This species lives in wood, scrub and marsh as well as hedgebanks.

VOLES AND LEMMINGS Muridae; subfamily: Microtinae (*110 species*)

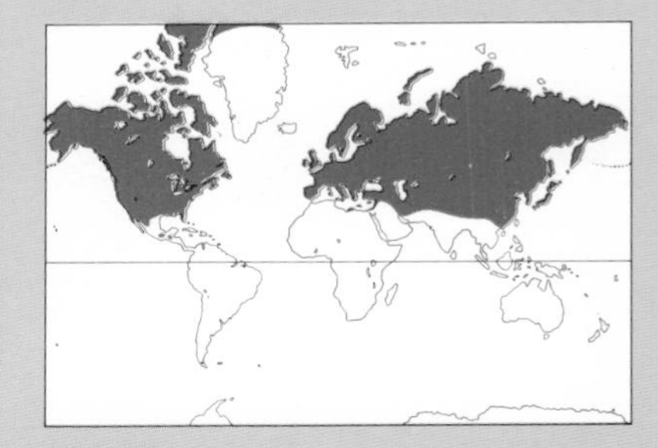

Habitat: tundra, scrub, open forest, grassland, rocky areas.

Diet: plants; a few species eat small insects and molluscs.

Breeding: average 3 litters per year, litters of 3-7.

Size: small voles: head-body 3in, tail 1in, weight ½ ounce; largest (muskrat): head-body 14in, tail 11in, weight 3lb.

Color: fawn, brown and gray to almost black.

Lifespan: a few months to 2 years.

Species mentioned in text:
Bank vole (*Clethrionomys glareolus*)
Collared or Arctic lemming (*Dicrostonyx torquatus*)
European water vole (*Arvicola terrestris*)
Muskrat (*Ondata zibethicus*)
Norway lemming (*Lemmus lemmus*)
Prairie vole (*Microtus ochrogaster*)

Voles and lemmings are small, stocky rodents. They have thick fur, short legs, blunt noses, small eyes and ears, and tails usually less than half the length of the body. They have many similarities with their cousins, the mice and rats. They too are mainly plant-eaters. They can breed at a great rate when conditions are good. And they live in all types of habitat.

▲The European water vole is at home in streams and lakes. Here one searches a river bank for food.

NORTHERN HOMES

However, unlike mice and rats, voles and lemmings live only in the Northern hemisphere. They do not inhabit tropical areas, nor are they nocturnal. Most species are active in bursts through the day and night, with brief rest periods between. This makes them vulnerable to a host of daytime and nocturnal predators, such as weasels, stoats, foxes, snakes, cats, owls, skuas and ravens.

Lemmings live in areas of coniferous forest, moor and the treeless, bleak tundra of the far north. Some vole species are found here too, but most live farther south, in temperate woodland, scrub and grassland. The mole-voles of central Asia are specialized for a burrowing life, with no visible ears and a cylindrical body. Their front teeth stick out forwards and are used for digging.

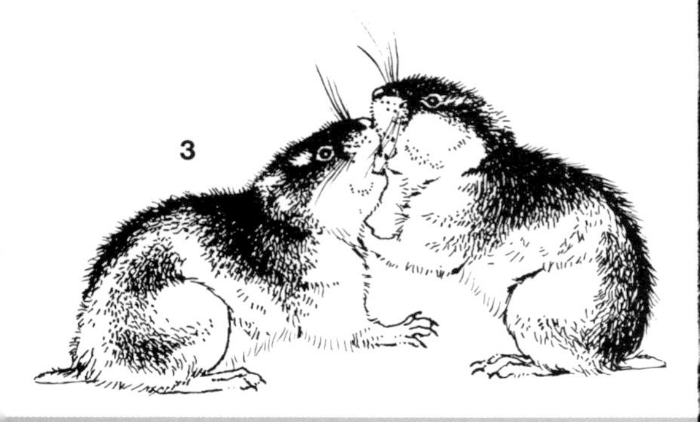

◀Norway lemmings, like other lemmings and voles, usually live alone. When two individuals meet, they may be aggressive and defend their territories. Here two males box (1), wrestle (2) and threaten(3).

▼During a mass migration, many lemmings die. This is not intentional suicide, but accidental drowning.

LONERS AND COLONIES

Most voles and lemmings live alone. They mark their territories using body secretions, and the odors warn others to keep out. Each animal defends its territory, especially at breeding time, by behaving aggressively, calling and even fighting intruders if necessary.

A few species, such as some meadow voles of North America, live in colonies. They dig complicated tunnels and keep in touch using a variety of squeaks, whistles and chirrups.

As with most rodents, young voles and lemmings are born blind and helpless. The female usually gives birth in a nest of dry grass, lichen or other vegetation, in a burrow. The average number in a litter varies from 2 to 12, depending on species, food supply and other conditions.

Parental care is the work of the mother, who guards her young carefully. If they wander from the nest, she finds them by their squeaks and brings them back. Only in the Prairie vole does the male help, grooming the young and keeping the runways and burrows clean.

TUNNELS UNDER THE SNOW

Lemmings are the most numerous small mammals in the tundra. During the short summer they feed on grasses, herbs and other plants that grow in the swampy soil. During the long, cold winter, they remain active by digging tunnels at ground level under the snow and feeding on roots, shoots, mosses, lichens and other plant matter. The blanket of snow is a good insulator. Bitter winds above may lower the air temperature to minus 40°F, but in the snow burrows it rarely falls much below freezing.

Norway lemmings live in the tundra of Scandinavia and the north-western USSR. In winter they stay under the snow, tunnelling, feeding and breeding in snug underground nests. With the spring thaw, these lemmings avoid floods by moving to higher ground or to more wooded areas. Here they make shallow tunnels in the soil, travel by well-marked runways through the undergrowth, and continue to feed and breed.

POPULATION EXPLOSION

The ability of the Norway lemming (and many other lemming species) to breed almost all year round means that its numbers can increase dramatically. Females mate before they are a month old, males soon after. With room to establish a territory and plentiful food, a female can have one litter each month. There is an average of six young per litter.

▲A Collared lemming cares for her young in their grassy nest on the edge of the treeless Arctic tundra.

In some years, the Norway lemming population increases very quickly. It may rise from less than 5 animals per 2½ acres (about the area of two football fields) to more than 300 animals. This happens roughly every 3 or 4 years.

The result is overcrowding, with not enough food to go around and too many encounters between individuals as each forages over its shrinking territory. Gradually some of the extra lemmings are forced to seek better conditions. This is the famous "lemming migration."

ON THE MARCH

Many tales surround the mass marches of lemmings. They happen in several species, not only in the Norway lemming. Thousands of animals descend from uplands and woods into the lowland fields, destroying crops and polluting rivers and wells with their dead bodies.

◀**Types of vole and lemming** A muskrat sits on its house of branches and twigs **(1)**. A tree-living species, the Red-tree vole (*Phenacomys longicaudatus*) **(2)**. A Norway lemming sniffs the air **(3)**. A Collared lemming in its white winter coat**(4)**, and in its brown summer fur **(5)**. A Taiga vole (*Microtus xanothognathus*) on its hind legs keeps watch **(6)**. A Meadow vole (*Microtus pennsylvanicus*) drums an alarm signal with its hind foot **(7)**. A Southern mole-vole (*Ellobius fuscocapillus*) **(8)**.

But these great marches are not deliberate suicide – despite what people once believed. Barriers such as mountains and rivers cause the travelling lemmings gradually to collect in enormous groups. When they reach the shore of a lake or the sea, the only answer is to try to swim across. Normally, lemmings are reasonable swimmers and attempt only short trips in the water. But on migration they seem to get caught in mass panic, unable to use normal judgement. They set off to swim across a sea or clamber down a vertical cliff. As a result, they die in their efforts to find new, unoccupied lands.

Lemming population explosions affect other animals. Arctic foxes, one of their chief predators, also become more numerous as the lemmings increase and the foxes have more food. Trappers in the fur trade realized that in some years they caught few foxes, then 3 or 4 years later the numbers would be very high, only to fall suddenly once again. Records of the Hudson Bay Company from 1870 to 1920 show fox pelt numbers varying from 200 to 6,000 in such a cycle.

WATERY LIFE

The muskrat is a very large vole, specialized for an aquatic life. Its name comes from the strong smell of the secretions made in two glands at the base of its tail. The muskrat's hind feet are partly webbed and have a fringe of stiff hairs, to aid swimming. Its tail is almost hairless, scaly and flattened from side to side – it works as a rudder in the water. The outer protective fur is coarse and long, varying in color from gray-brown to black. The underfur is short, thick and very tough. It is sold by traders as the valuable "musquash," which is the name given to the animal by the Canadian Plains Indians.

Muskrats live in North America, in swamps, rivers and lakes. They eat water plants and also some animals such as mussels, crayfish, snails and fish. They build nests in grassy banks and dig elaborate tunnels with entrances that are often under water.

These voles were taken to Europe, to breed on fur farms. They escaped and now live wild in certain areas, such as the Netherlands. They can do great damage, tunnelling into dykes and dams and eating crops. In temperate climates they can breed from spring to fall and raise several families each year, and so quickly become pests.

Also, in Europe, in the absence of many of their natural predators – coyotes, raccoons, turtles, water snakes and alligators – muskrats can spread far and wide. However, in other places in the world they have been released on purpose, since their feeding prevents waterways from becoming choked by weeds.

◀The European water vole's glossy brown outer coat hides the short, dense underfur that keeps the animal warm and dry.

HAMSTERS

As spring warms the frozen ground of the Russian steppes, hamsters awaken from their winter hibernation and emerge in search of food. Because food is often scarce, they must search a wide area. Each hamster jealously guards its own territory. Two hamsters meet along the edge of their territories and begin to fight one another. One of them receives deep wounds and runs away. These fierce little creatures will attack animals ten times their own size.

◄The dwarf Dzungarian hamster is not much bigger than a fir cone.

Hamsters are rodents related to rats and mice. The familiar Golden hamster, often kept as a pet, has a squat, rounded body, blunt face, large ears and a tiny tail. Other hamsters are more mouse-like, with a pointed snout and a longer tail.

Hamsters are solitary animals, and each one digs its own underground burrow a few feet below the surface. The Common hamster builds a warren with several entrances and many chambers. One chamber may be used for sleeping, a second as a lavatory and others as food stores.

▲The Common hamster's cheek pouches consist of loose folds of skin. The pouches expand as food is pushed in.

FOOD HOARDS

The hamsters' diet consists mainly of plant material, including roots, seeds and leaves, but they also eat insects and the occasional snake or young bird. They carry food in their cheek pouches. The front paws are used like hands to pack seeds and nuts into these elastic folds of skin. When food is plentiful, hamsters gather it into their cheek pouches and in this way carry it underground to their burrows. Stored there, it lasts throughout the barren winter months. The storage chamber of a single Common hamster was found to contain 200lb of food.

FAST BREEDING

When they are kept in artificially warm conditions in cages, Golden hamsters produce young at any time of the year. In the wild they normally breed just twice, in spring and summer. Mating adults make high-pitched squeaks which cannot be picked up by the human ear. Two weeks after mating, the blind naked young are born in a nest lined with soft wool and grasses. The male plays no part in raising his offspring. The female suckles them for about 3 weeks.

The young are able to mate at the age of about 2 or 3 months. Common hamsters mature even earlier, at about 6 weeks. Given plenty of food, they have been known to breed even before they leave the nest.

Individual hamsters only live for 2 or 3 years in the wild, but because they reproduce quickly their numbers remain high. All of the millions of Golden hamsters in Western Europe and North America have been bred from a group of 13 originally collected in Syria in 1930.

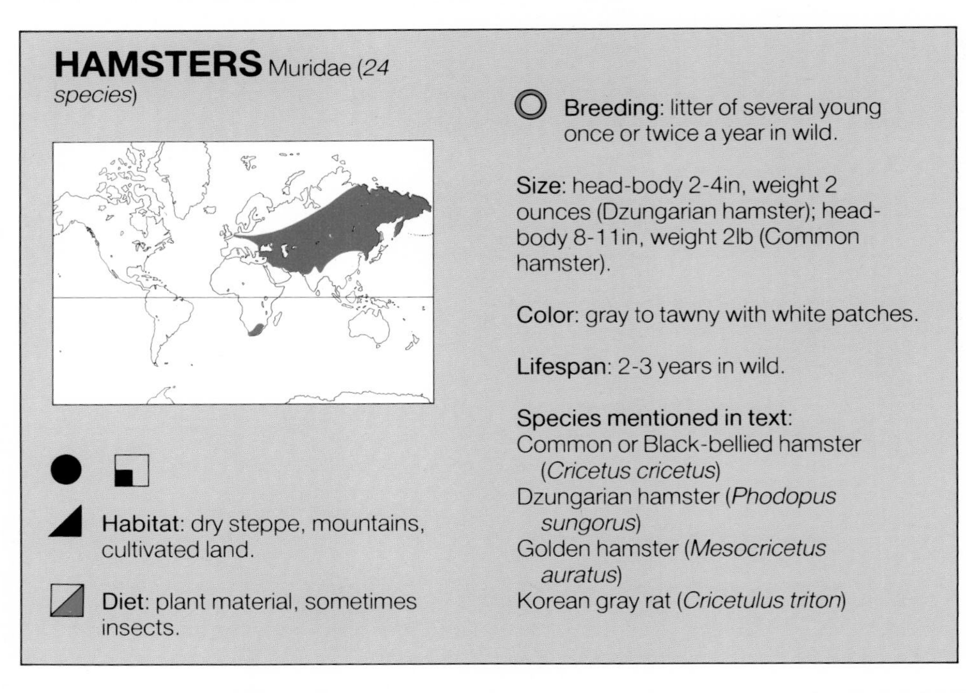

HAMSTERS Muridae (*24 species*)

Habitat: dry steppe, mountains, cultivated land.

Diet: plant material, sometimes insects.

Breeding: litter of several young once or twice a year in wild.

Size: head-body 2-4in, weight 2 ounces (Dzungarian hamster); head-body 8-11in, weight 2lb (Common hamster).

Color: gray to tawny with white patches.

Lifespan: 2-3 years in wild.

Species mentioned in text:
Common or Black-bellied hamster (*Cricetus cricetus*)
Dzungarian hamster (*Phodopus sungorus*)
Golden hamster (*Mesocricetus auratus*)
Korean gray rat (*Cricetulus triton*)

▲The Common hamster has large mobile ears for picking up the smallest sounds. The front paws are used like hands for holding food.

ENEMIES AND PREDATORS

In many places hamsters are treated as pests because they damage crops. They are particularly partial to wheat, barley, millet, soy beans, peas, potatoes, carrots and beets. They are also trapped for their skins and for food. Happily, they mainly occur in cold, barren places where people do not live. Their main enemies are wolves and polecats, which can dig them out of burrows with their powerful claws.

The Korean gray rat, which is a hamster despite its name, rolls over on to its back and shrieks if attacked. This shock tactic may give it just enough time to escape from a very surprised predator.

GERBILS

Quietly and cautiously, a gerbil peeps out of its burrow. It is dusk on the dry Asian grassland, and cool enough to begin the nightly search for food. This small, mouse-like rodent begins to creep along a bank, beady-eyed and whiskers quivering, sniffing out seeds.

GERBILS Gerbillinae (*81 species*)

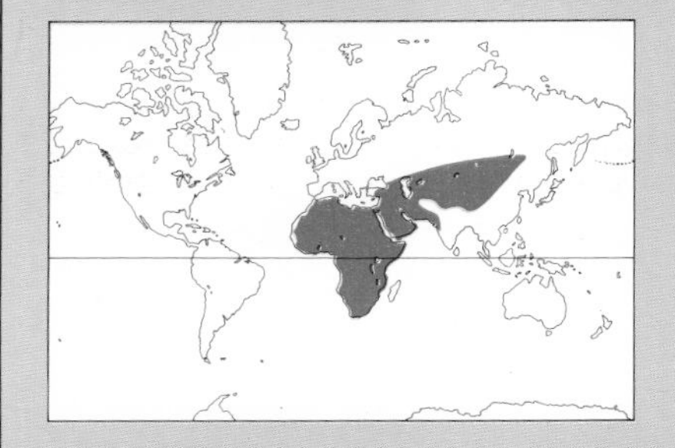

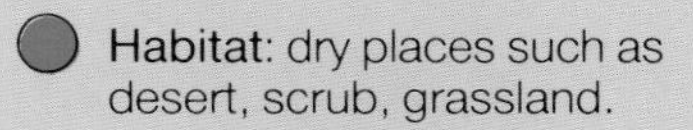

Habitat: dry places such as desert, scrub, grassland.

Diet: seeds and other plant food; sometimes insects and other small animals.

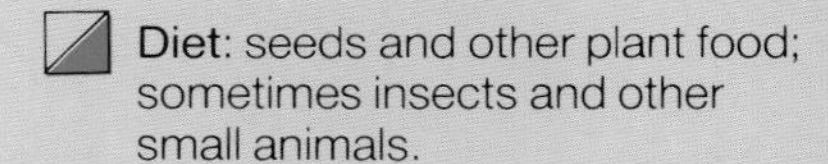

Breeding: in driest areas 1 litter after rainy season, in damper places up to 3 litters through year, average 3-5 young per litter.

Size: head-body 3-8in; tail 3-9in; weight ¼-7 ounces.

Color: pale gray, reddish or brown to almost black; varies within species to match color of surroundings.

Lifespan: 1-2 years.

Species mentioned in text:
Great gerbil (*Rhombomys opimus*)
Mongolian gerbil (*Meriones unguiculatus*)
South African pygmy gerbil (*Gerbillurus paeba*)
Wagner's gerbil (*Gerbillus dasyurus*)

Many people recognize the gerbil (or jerbil as it is sometimes spelt), since it is often kept as a pet. But there is not just one species of gerbil – there are about 80. They live in Africa, the Middle East and across central Asia to India and China. The familiar pet is usually a Mongolian gerbil.

Gerbils are cousins of rats and mice, which they resemble except for their furry tails. Some species are called rats or sand rats, and others are called jirds or dipodils, but they all belong to the gerbil group.

Gerbils are specialized for life in dry places such as desert, dry savannah, steppe grassland and rocky areas. Those from very dry places usually live alone. In scrub or grassland, where food is more plentiful, they may live in pairs or groups of up to 20.

To avoid the heat and being seen by predators, gerbils usually hide by day in their burrows. These are about 20in deep, and the air in them is cool and moist all the time. To keep the burrow cool, some gerbils block its entrance with stones or bits of plants. At night, when it is not so hot, they emerge under cover of darkness for a night-time feast.

SEEDS AS SPONGES

Gerbils eat mainly seeds and other parts of plants such as roots, shoots, buds, leaves and fruits. But in such harsh surroundings, any food is welcome. Some gerbil species eat insects, lizards, worms and other small creatures, as well as plants. Wagner's gerbil eats snails, leaving big piles of empty shells outside its burrow.

All animals need water to survive. In the gerbil's dry surroundings, there is little water to drink, and the food is often dry too. The gerbil is an expert at making the best of what is available. When it starts to feed, the dew is falling and the seeds soak up some of this moisture, like tiny sponges. To save more moisture, gerbils produce no more than a few tiny drops of urine each day.

Gerbils often carry food back to the burrow, to eat in safety or store until later. In the damp burrow, the food absorbs yet more moisture. The Great gerbil of central Asia even stores food outside its burrow, in piles more than 3ft high!

ATTACKS FROM THE AIR

Many gerbils fall victim to flying predators, such as owls and hawks. But most species are well adapted to reduce the risk of being captured. They are colored like their surroundings, and have a keen sense of hearing.

▲Like most gerbils, the South African pygmy gerbil is a hoarder. It stores food in its burrow, to eat when times are hard.

◀**Busy about their business** A South African pygmy gerbil **(1)** cleans its whiskers. A Tamarisk gerbil (*Meriones tamariscinus*) **(2)** looks for danger. A Libyan jird (*Meriones libycus*) **(3)** leaps at an enemy, while a Cape short-eared gerbil (*Desmondillus auricularis*) **(4)** crouches in submission. A Great gerbil **(5)**, one of the largest species, marks its territory with a pile of sand, urine and droppings. The small *Gerbillus gerbillus* **(6)** does the same with fluid from a gland on its underside. A female Mongolian gerbil **(7)** raises her neck-fur during courtship. A Fat sand rat (*Psammomys obesus*) **(8)** sniffs a ball of sand and urine left by a neighbor.

DORMICE

The cold October wind sweeps through a small European town. For several days a dormouse has been making a snug nest of dry grass and moss in a thick hedge bordering a garden. It is a tubby animal, yet it still climbs nimbly through the bushes as it feeds on berries and nuts. Eventually the dormouse settles to sleep in the nest. When it wakes, it will be the middle of spring.

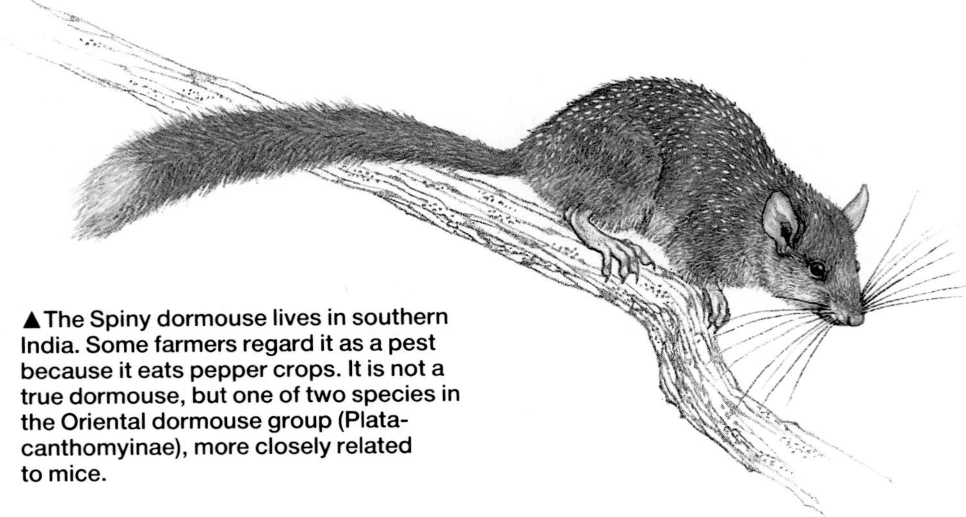

▲The Spiny dormouse lives in southern India. Some farmers regard it as a pest because it eats pepper crops. It is not a true dormouse, but one of two species in the Oriental dormouse group (Platacanthomyinae), more closely related to mice.

►The Garden dormouse, like its relatives, has sharp, curved claws and is an expert climber.

The dormouse has nothing to do with doors. Its English name comes from the French word *dormir*, meaning "to sleep." And in Europe this is what a dormouse does. For about 7 months every year it sleeps, non-stop, in deep hibernation through the winter.

During fall the dormouse feeds greedily on fruits, nuts and berries, building up a layer of body fat. The fat will provide energy for the long winter sleep. As it goes into hibernation, the dormouse's body temperature falls. Its breathing and heartbeat slow down. It will not wake up, even if one prods or pokes it, like the dormouse at the tea-party in *Alice in Wonderland*. Hibernation is one way of surviving the winter, when there is little food, and the freezing temperatures are dangerous for many small animals.

CALLING AND WHISTLING

Around April, warmer weather wakes the dormouse from its slumber. Thin again, it feeds hungrily on buds, shoots, flowers and small creatures. It is now the mating season, and dormice make a variety of noises at this time. The female Garden dormouse whistles to attract her mate. The male Edible dormouse makes a squeaky call as he chases a female.

Edible dormice are so named because the Romans used to keep them in cages and fatten them for food. One of the world's first cookbooks, written by Epicius of Ancient Rome, has a recipe for stuffed dormouse. The Edible dormouse has a very bushy tail and looks like a Gray squirrel. Most other species have rusty red or brown fur and look like a cross between a squirrel and a mouse.

NURSERY NEST

Around 3 to 4 weeks after mating, the female dormouse builds a ball-shaped nest in a tree hole or the crook of a branch. She lines it with feathers and hairs. Here she gives birth to about four blind, furless, helpless young. After 3 weeks the young can see and hear. By 6 weeks they are fending for themselves. They are able to breed when about a year old, towards the end of their first hibernation.

Most species of dormouse live in loose groups and do not stray from their home range. Locally they are quite common. The Garden dormouse has taken advantage of human presence and lives in gardens in towns and villages, although it also inhabits forests and rocky places.

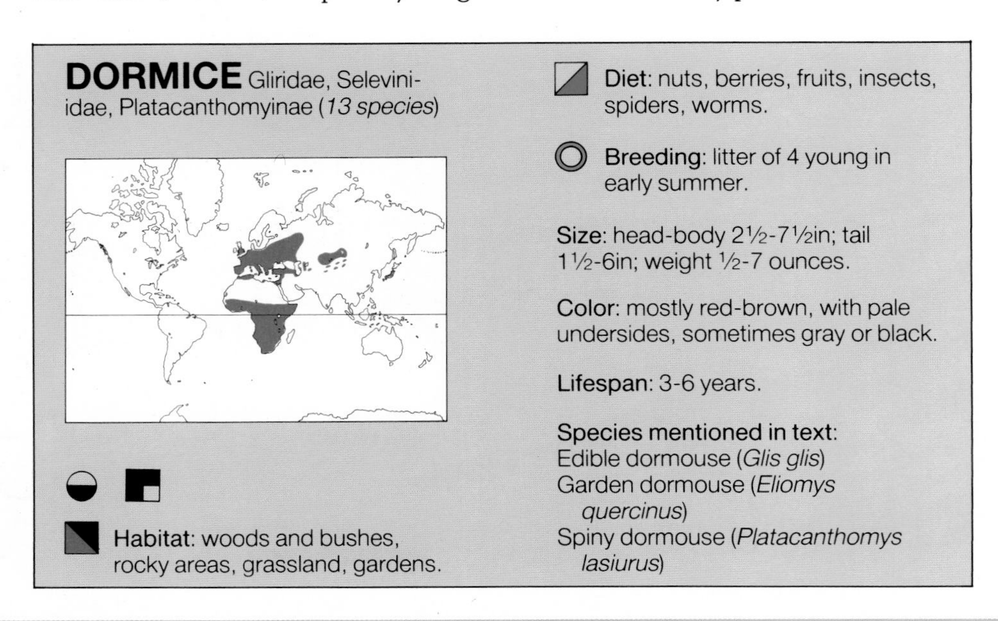

DORMICE Gliridae, Seleviniidae, Platacanthomyinae (*13 species*)

Habitat: woods and bushes, rocky areas, grassland, gardens.

Diet: nuts, berries, fruits, insects, spiders, worms.

Breeding: litter of 4 young in early summer.

Size: head-body 2½-7½in; tail 1½-6in; weight ½-7 ounces.

Color: mostly red-brown, with pale undersides, sometimes gray or black.

Lifespan: 3-6 years.

Species mentioned in text:
Edible dormouse (*Glis glis*)
Garden dormouse (*Eliomys quercinus*)
Spiny dormouse (*Platacanthomys lasiurus*)

JERBOAS

In the harsh Saharan sunset, the light and heat fade, as three jerboas hop from a burrow and begin the night's foraging. But one starts to stray, feeding on a wind-blown trail of seeds. A desert fox, watching nearby, creeps quietly between this animal and its burrow. The jerboa's large ears detect the faint sound of the fox padding across the sand. With two huge leaps the jerboa tries to escape, but the fox is too quick. All that's left of the jerboa are entrails on a blood-stained patch of sand.

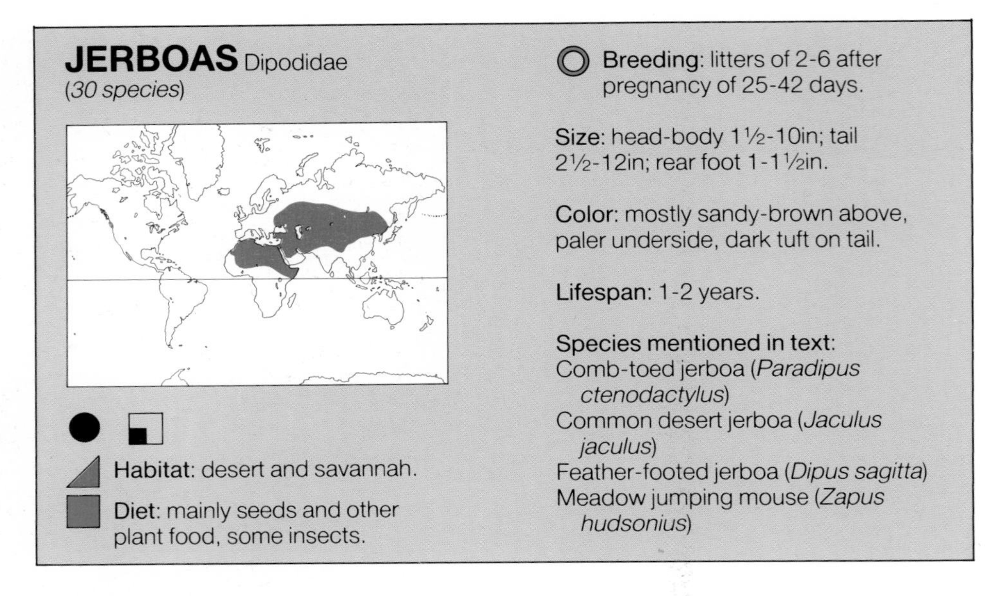

JERBOAS Dipodidae
(*30 species*)

Habitat: desert and savannah.

Diet: mainly seeds and other plant food, some insects.

Breeding: litters of 2-6 after pregnancy of 25-42 days.

Size: head-body 1½-10in; tail 2½-12in; rear foot 1-1½in.

Color: mostly sandy-brown above, paler underside, dark tuft on tail.

Lifespan: 1-2 years.

Species mentioned in text:
Comb-toed jerboa (*Paradipus ctenodactylus*)
Common desert jerboa (*Jaculus jaculus*)
Feather-footed jerboa (*Dipus sagitta*)
Meadow jumping mouse (*Zapus hudsonius*)

▲A desert jerboa scrabbles in the sand with its short forelegs, looking for seeds, shoots and other tidbits.

▼A jerboa uses its long tail to balance when leaping, and also as a prop to lean on when standing still. These small rodents hop distances many times their own body size – a desert jerboa can jump 10ft with each leap.

The jerboa's body design, with its enormously long hind legs, shows at once that this animal is built for jumping. The hind legs are at least four times longer than the front ones, and the strong rod-like bones act as levers to propel the animal forwards in great bounds.

Jerboas do not leap everywhere, however. When not in a hurry they can hop slowly, or even walk, using only their hind legs.

HAIRY FEET

Species of jerboa living in sandy areas have tufts of fur on the undersides of their feet. These act as "snowshoes," preventing them from sinking into the soft sand, and they also help gain a grip when taking off on a leap. The Feather-footed jerboa of China and the USSR and the Comb-toed jerboa of the USSR are named after their furry feet. Hairy feet are useful too when digging for food or making a burrow. The jerboa scrapes away soil using its forelegs and its rodent's large front teeth, and kicks away the soil behind with its powerful hind legs.

HOMES FROM HOME

Most species of jerboa live in hot deserts, sandy scrubland and other dry places, across North Africa and central Asia. Since there are few trees or bushes in which to hide, jerboas dig burrows. Here they stay during the day, away from the heat and the drying effect of the Sun. At night they emerge to search for food such as seeds, leaves and shoots and also beetles and other insects. Their large eyes help them see in the dark, and their big ears detect faint sounds.

The Common desert jerboa lives in loose groups, but most other species

live alone. Each jerboa may have several burrows in the area where it lives. There is usually one main burrow, which is used most of the time, and where a female gives birth to her young. This may be as deep as 7ft and have several connecting chambers and passages, plus "larders" for food storage. There is often a spare exit for emergencies, or a chamber that is just under the surface, from where the jerboa can "burst" out of the ground and make its escape.

There are also smaller burrows, from 4 to 20in long, which the jerboa uses during the day or at night if it cannot return to its main burrow.

In the more northern parts of their range, jerboas also have hibernating burrows. Here they sleep through the cold winter. In spring they mate, and the young are born in a nest deep in the nursery burrow. In about 2 months they are ready to fend for themselves. Jerboas average two litters a year.

Related to the jerboas, and with the same large hind feet for hopping, are the jumping mice (family Zapodidae). There are 14 species, living across central North America, Eastern Europe and central Asia.

▲The Meadow jumping mouse is a relative of the jerboas. It has the same hopping habits, but it does not burrow.

▼A desert jerboa's large ears listen for predators. They also lose heat to the surroundings, keeping the animal cool.

PORCUPINES

Grunting quietly to itself, an African porcupine is digging up roots for its evening feed. It is so busy that it does not hear the soft padding footsteps of the young leopard closing in behind it. The cat pounces, but instead of an easy meal it gets the shock of its life. With a snort of rage the porcupine raises its quills and charges backwards. Almost too late the leopard sees the danger. As it throws itself to one side, the needle-sharp quills miss its face by less than 3ft.

All 22 species of porcupine have quills and spines. They are more obvious in some species than in others, and they come in many shapes and sizes. These quills and spines are highly modified hairs, providing their owners with a very effective defence.

PORCUPINES Hystricidae and Erethizontidae (*22 species*)

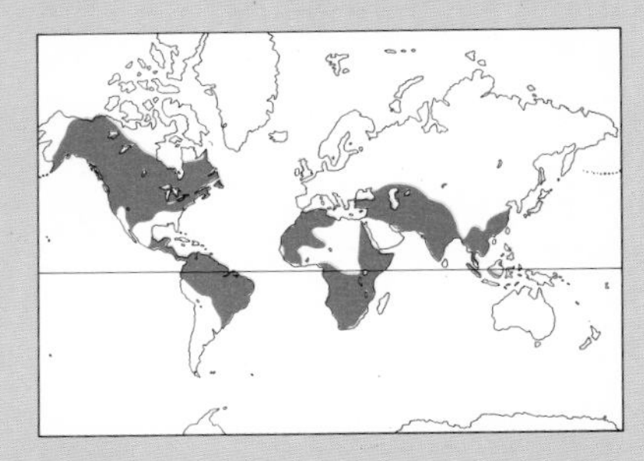

Habitat: forest, open grassland, desert, rocky areas.

Diet: leaves, roots, fruits.

Breeding: 1 or 2 young after pregnancy of about 210 days (North American porcupine), 93 days (Cape), 112 days (African).

Size: smallest (Prehensile-tailed porcupine): head-body 12in, weight 2lb; largest (crested porcupines): head-body 34in, weight 60lb.

Color: from black, white and gray, through yellow-brown to gray-brown.

Lifespan: up to 17 years.

Species mentioned in text:
African porcupine (*Hystrix cristata*)
Brush-tailed porcupines (*Atherurus* species)
Cape porcupine (*Hystrix africaeaustralis*)
Crested porcupines (*Hystrix* species)
Indonesian porcupines (*Thecarus* species)
North American porcupine (*Erethizon dorsatum*)
Prehensile-tailed porcupine (*Coendon prehensilis*)
South American tree porcupine (*C. bicolor*)

SPECTACULAR QUILLS

The most spectacular of the specialized hairs are the 20in black-and-white quills of the crested porcupines of Africa and Asia. They cover the animal's back from the shoulders to the tail, and they can be raised by muscles under the skin. They cannot be "shot" at attackers but they often stick in the face, neck or paws of an attacker and break off. They are not poisonous, but can become infected and so can cause serious injury.

The crested and brush-tailed porcupines of Africa and Asia also have clusters of tail quills modified into rattles. When the animal is frightened or cornered it shakes the tail to give a warning rattle. If that does not work, the African porcupine stamps its hind feet and grunts, and as a last resort charges its attacker.

The protective quills of the North American porcupine are different. They are only about 1in long and are hidden in the animal's long coarse hair. They are very loosely fixed in the porcupine's skin, and an attacker is likely to find itself with nothing but a mouthful of hair and spines. To make things even more painful, the tips of the quills are covered in tiny barbs.

▼A Cape porcupine feeding on desert gourds. Most porcupines feed in this way, holding the food with their front paws and nibbling at it.

►A South American tree porcupine caught on the ground as it crosses from one tree to another. Just visible in the grass is the bare patch on the upper surface of the end of the tail. This patch of hard skin (the callus) helps to improve the grip of the prehensile tail.

▼**Distant cousins** Three members of the porcupine family from widely separated parts of the world. The North American porcupine **(1)** spends most of its time on the ground. The Indonesian porcupine **(2)** has a dense coat of flat, flexible spines. There are three species – in Borneo, Sumatra and the Philippines. The African porcupine **(3)** is one of five crested species. It is very adaptable and is found in desert, grassland and forest.

WORLDS APART

The porcupines are grouped in two large families. The Old World porcupines live in a wide range of habitats in Africa and Asia. New World porcupines are found in forest and grassland areas from northern Canada down to northern Argentina. The two groups have similar quills, teeth and jaw muscles. But scientists have still not worked out whether they are descendants of the same ancestors or whether evolution has come up with the same design in two separate families.

"OLD WORLD" SPECIES

All the Old World porcupines are ground-dwelling animals. They feed on roots, bulbs, fruits and berries, usually at night, either alone or in pairs. During the day they rest in caves, in holes among rocks or in burrows. Sometimes they take over old aardvark holes.

In many parts of Africa, porcupines are hunted for their arguably tasty meat. They are also killed by farmers because of their habit of raiding crops of maize, groundnuts, melons and potatoes. In spite of this, the porcupine is still common in Africa, probably because its natural enemies – lions, leopards and hyenas – are themselves becoming rare.

The sharp quills of the crested porcupines make mating rather dangerous, but somehow the animals manage. The young are born in a grass-lined chamber in a burrow or rock shelter, usually in summer. The babies are born with their eyes open. They are covered with fur and even have tiny soft quills which harden within a few hours.

The Cape porcupine of South Africa lives in family groups of up to eight. The older animals help to look after the babies, keeping close to young porcupines when they first start to feed outside the burrow at about 6 to 7 months.

IN THE AMERICAS

Unlike their Old World relatives, the porcupines of North and South America are mainly tree-dwellers. Their feet are equipped with large claws and hairless pads that provide a good grip. Most specialized of all are the Prehensile-tailed porcupine and the tree porcupines that live in the Central and South American forests. They have long muscular tails, which they can coil round branches to give extra support when climbing.

These animals too are nocturnal. They feed mainly on leaves, but also take fruits and seeds and sometimes come down to the ground to feed on roots and tubers. In winter, the North American porcupine feeds on pine needles, leaves and tree bark. It is especially fond of Red spruce and Sugar maple. In summer it often feeds on grasses and on roots, berries and flowers on the forest floor.

Destruction of South America's forests has placed several species in danger. The Prehensile-tailed porcupine is threatened in parts of Brazil. And the South American tree porcupine is listed as endangered by the Brazilian Academy of Science.

◀The Prehensile-tailed porcupine spends most of its time high in the trees. It is near-sighted, relying heavily on touch and smell.

▶Despite its slowness, a hungry North American porcupine will sometimes climb 60-70ft to reach young leaves.

CAVIES

Half hidden by a thorn-bush a female mara or Patagonian hare scans the dry Argentinian scrubland for danger. She is an odd-looking animal, with the big upright ears of a hare and the oval body and slender legs of a small antelope. Slowly she approaches the dark mouth of a den dug in the side of a dry earth bank. She pauses and lets out a shrill whistling call. In a shower of dust, a dozen young maras burst into sight. Shoving and pushing, they almost knock the female over in their mad scramble to get at her milk. But not all the young are hers, and she is not going to be bullied. Pushing the rest aside she selects her own two babies and leads them away for their morning feed. Nearby her mate stands guard.

The mara is closely related to the guinea-pig and the Rock cavy, but is unlike the rest of the cavy family. It looks different, and its life-style is different. Uniquely among cavies, maras pair for life. Male and female spend all their time together and for most of the year avoid other maras.

CAVIES Caviidae
(14 species)

Habitat: grassland to desert.

Diet: grasses, herbs, leaves.

Breeding: desert cavies: 2-4 young after pregnancy of 50 days; Rock cavy: 1-2 young after 75 days; mara: 1-3 young after 90 days.

Size: cavies: head-body 9-15in, weight 2/3-2lb; mara: head-body 20-30in, tail 1½in, weight 17½-20lb.

Color: white, gray, yellow-brown to reddish-brown, wild species duller.

Lifespan: cavies 4 years; mara 15.

Species mentioned in text:
Domestic guinea-pig (*Cavia porcellus*)
Mara (*Dolichotis patagonum*)
Rock cavy (*Kerodon rupestris*)

In the breeding season as many as 15 pairs may gather in one small area. As soon as their young are born they are placed in a communal den dug by the females. Young maras are able to move around and nibble grass within 24 hours of birth, but until they are 4 months old they are fed once or twice a day by their mothers. Each female visits the den in turn, while her mate stands guard over their young and keeps other adults away.

Near the end of the breeding season maras sometimes gather in groups of up to 100 on the partly dried out beds of small lakes. Here the moist ground provides a short but welcome feast of grass and herbs.

MOST "TALKATIVE" RODENTS

The most familiar cavy is the domestic guinea-pig. It is no longer found anywhere in the wild, but its 11 close relatives are among the most successful and abundant rodents in South America. They are all built to the same body-plan with a short thick body, large head and sharply clawed feet.

▼Even when maras gather in large groups, the male-female pairs remain faithful to each other. Males keep a close watch on their mates and chase other males away.

They are also among the most "talkative" of rodents and have a great variety of chirps, squeaks, grunts and whistles. Some species also drum on the ground with their back feet as a warning of danger.

VEGETARIANS

The cavies are all vegetarians, and between them they have adapted to just about every kind of open habitat that South America has to offer. For example, members of the genus *Cavia* generally prefer moist grasslands and feed on grasses and herbs. Those of the genus *Microcavia* inhabit the Patagonian desert and the dry *puna* region of the Andes. They climb well and feed mainly on the leaves of trees and bushes. One of the most specialized is the Rock cavy, which lives in rocky outcrops dotted over the dry thorn-scrub of north-east Brazil.

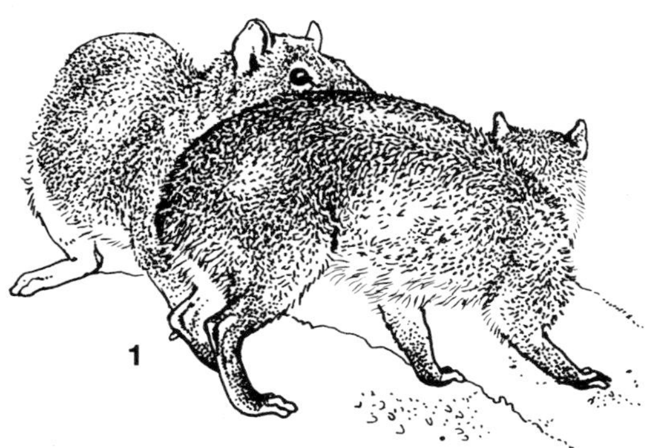

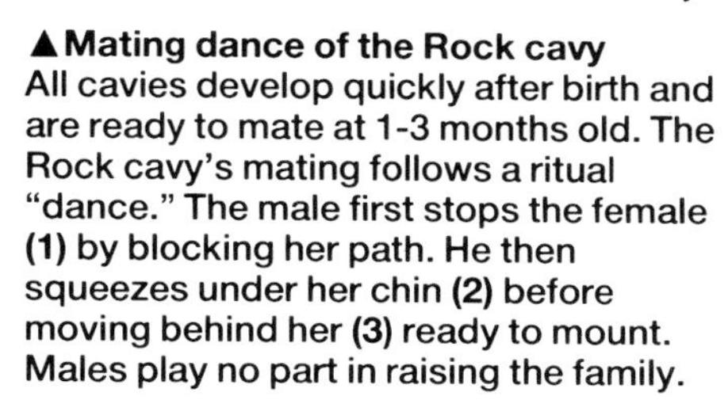

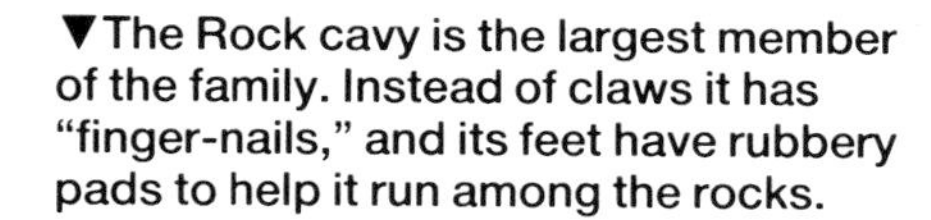

▲Mating dance of the Rock cavy
All cavies develop quickly after birth and are ready to mate at 1-3 months old. The Rock cavy's mating follows a ritual "dance." The male first stops the female **(1)** by blocking her path. He then squeezes under her chin **(2)** before moving behind her **(3)** ready to mount. Males play no part in raising the family.

▼The Rock cavy is the largest member of the family. Instead of claws it has "finger-nails," and its feet have rubbery pads to help it run among the rocks.

CAPYBARA

The sight of a herd of capybara feeding quietly on the banks of the River Amazon is like a glimpse into prehistoric times. With their massive square heads and heavy bodies they look more like the primitive animals of 50 million years ago than modern members of the guinea-pig family. Yet that is exactly what they are. Members of the group keep in touch with one another with throaty purring sounds.

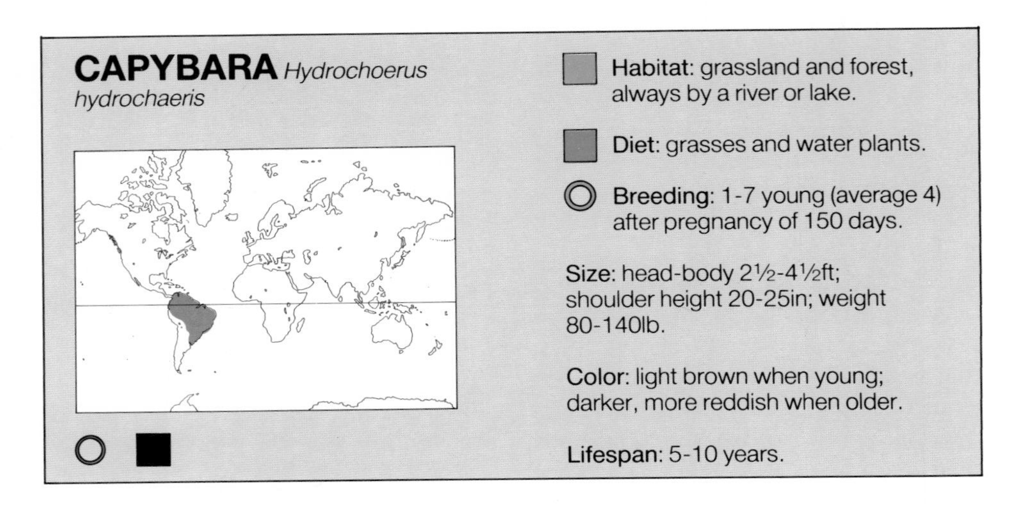

CAPYBARA *Hydrochoerus hydrochaeris*

Habitat: grassland and forest, always by a river or lake.

Diet: grasses and water plants.

Breeding: 1-7 young (average 4) after pregnancy of 150 days.

Size: head-body 2½-4½ft; shoulder height 20-25in; weight 80-140lb.

Color: light brown when young; darker, more reddish when older.

Lifespan: 5-10 years.

▲A male capybara leaving his tell-tale scent on a low branch by the river.

Capybaras are found only in South America, on the shores of rivers and lakes. They inhabit a wide variety of habitats, from open grassland to tropical rain forest. But no matter where they live, they are never very far from water.

They are big, heavy animals – the world's largest rodents – with barrel-shaped bodies and long shaggy coats. Their back legs are longer than their front legs, which gives them a lumbering, galloping run. Once in the water they are powerful swimmers. Their toes are partially webbed, and their eyes, ears and nostrils are high on the head so they remain clear of the water when swimming.

Capybaras head for the water at the first sign of danger. They can swim underwater for up to 5 minutes when trying to escape from a predator.

◀The large hairless bump on the male capybara's snout is called the "morrillo." It is a gland that produces a sticky white liquid used as a scent-marker.

▼The capybara's long back legs give it a rabbit-like shape when lying down.

ALL-NIGHT DINERS

Capybaras are complete vegetarians. They eat a variety of water plants, but their main food is grass. They have large chisel-like front teeth and are efficient at cropping the short dry grass that remains after the dry season.

Mornings are usually spent resting, but as the temperature rises around midday the animals take to the water. Early evening is the main feeding time, but even after dark the capybara seldom sleeps for long. All through the night the animals are active, alternating short naps with spells of leisurely feeding.

BOSS OF A BAND

Capybaras live in bands or herds which vary in size throughout the year. A typical group will have one "boss" male, two or three females and their young, and perhaps two or three junior adult males. There is a clear order of rank among the males. The leader constantly reminds the juniors of his rank by chasing them around and by herding the females and young together.

In the wet season, groups vary in size between 10 and 40, but in the dry season bands of over 100 may gather round waterholes or in places with good grazing.

The capybara's main enemies are jaguars and caimans, but young ones are often taken by foxes, semi-wild dogs, eagles and vultures.

Capybaras start to breed when they are about 18 months old. Mating always takes place in the water. At the end of the long pregnancy the female leaves her group for a while, and the young are born in the cover of a long grass thicket. Within a few days they can walk and within a week they are feeding on grass.

COYPU

In the spring of 1988, British government officials announced that there were no longer any wild coypus living in the country's waterways. For more than 50 years this large, shaggy, South American river animal had made its home in the marshes and fens of the Norfolk Broads. But its river-bank burrowing and visits to sugar-beet fields had made it many enemies. Eventually it was outlawed – and removed.

The coypu is a native of the rivers and marshes of central and southern parts of South America. It is a big, powerfully built rodent, which makes its home in a deep burrow tunnelled into the river bank. It is perfectly adapted for life in the water. Its coat is thick, waterproof and made of two layers. The long outer guard-hairs lock together when wet and help to trap a layer of air in the thick soft underfur beneath.

The coypu is an excellent swimmer, with fully webbed back feet. It feeds mainly on water plants, but often swims down to the river bed to add shellfish to its diet.

QUICK STARTERS

The coypu's young are born in a warm nest lined with grass, in a chamber at the end of the burrow. They are well covered with fur from the start and are born with their eyes open. The mother's milk is very rich, and the young grow quickly. Within a few days they can move about, and soon they are able to swim strongly.

Being able to move so soon after birth is very important. If the young were helpless for too long they would run the risk of drowning if the river level rose suddenly and water flooded into the nest chamber of the burrow.

The coypu is so specialized for life

COYPU *Myocastor coypus*

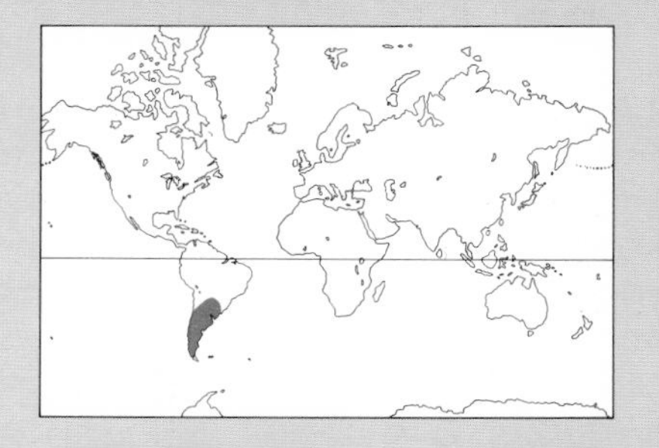

Habitat: alongside rivers and streams with high banks, also in salt-water inlets and marshes.

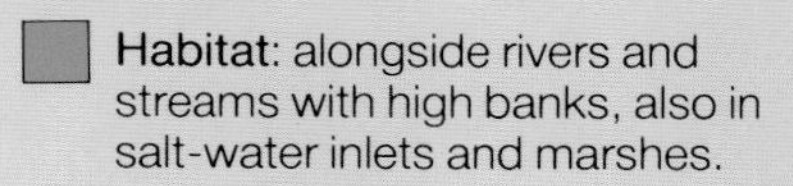

Diet: mainly water plants, with some shellfish; takes root crops (eg sugar-beet) where available.

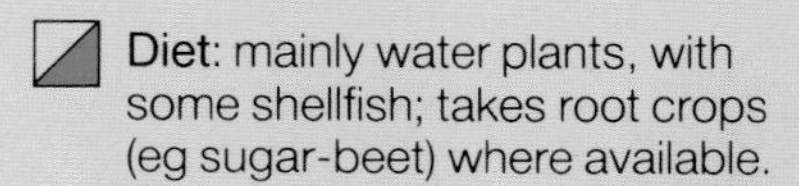

Breeding: litters of up to 9 after pregnancy of 120-150 days.

Size: head-body 17-25in; tail 10-18in; weight 15½-20lb.

Color: dark brown to yellow-brown, with gray or white muzzle.

Lifespan: not known.

►Wherever coypus have been farmed, some have escaped into the wild. This one was photographed in East Africa.

in the water that the mother can even suckle the young while swimming. The teats are placed high up on the sides of her body, just above the "water-line." There they can be seen and reached by young swimming alongside her or even by young riding on her back.

TOO SUCCESSFUL

In the 1920s and 1930s thousands of wild coypus were brought to Europe, the United States, Canada, Russia and Africa so that they could be bred for their valuable furs. Over the years some escaped. Others were released by their owners at times when business was bad. Wherever they were released, the coypus quickly made themselves at home. Soon they were more abundant in some of their new homes than in South America.

In Britain, for example, in the 1940s there were just a few coypus in one river in East Anglia. By 1972 the population had risen to about 12,000 and was rapidly becoming a problem. The animals' burrows were damaging river banks and ditches, and in some areas the coypus were also raiding crops such as sugar-beet. To stop the damage it was decided that the animals would have to be removed. Now there are no more coypu farms, and the wild coypus too are gone from the region.

▲The coypu's fur is darkish brown-yellow. The outer hair is long and thick, covering the soft underfur beneath.

▼Many rodents are good swimmers, but the coypu, with its webbed feet and thick waterproof fur, is a truly aquatic animal.

GUNDIS

With a stretch and a yawn the small furry animal sits up and looks around. For the past hour the gundi has been sunbathing on top of a boulder in the Algerian desert. Near by its parents and brothers and sisters are dozing. The animal sets about grooming its silky fur using the stiff bristles and horny combs of its toes.

GUNDIS Ctenodactylidae
(*5 species*)

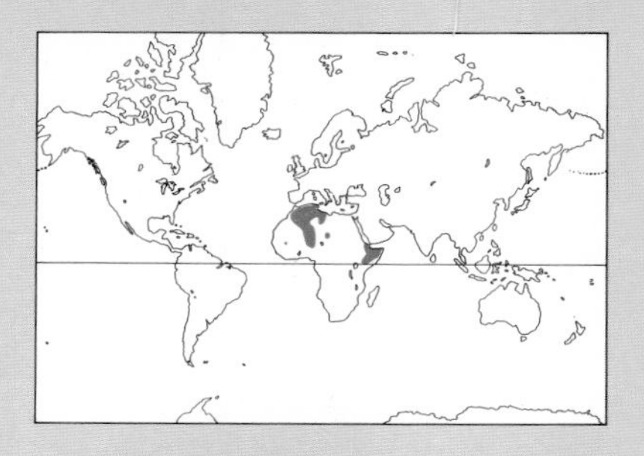

Habitat: rocky outcrops in desert, semi-desert and mountain areas.

Diet: grass, leaves and seeds.

Breeding: 2-4 young after pregnancy of 56 days (details known only for North African gundi).

Size: head-body 7in; tail 1-2in; weight 6-7 ounces.

Color: beige-brown, sandy to red-brown above, grayish-white underneath.

Lifespan: 3-4 years, up to 10 in captivity.

Species mentioned in text:
North African gundi (*Ctenodactylus gundi*)
Speke's gundi (*Pectinator spekei*)

The gundi's family name means "comb-fingers." Four species inhabit the hills and stony deserts on the northern edge of the Sahara Desert. The fifth, Speke's gundi, lives in the coastal hills and semi-desert regions of Ethiopia, Somalia and northern Kenya.

In size and shape all gundis are similar. They are very like guinea-pigs, and a full-grown adult would just fit on a large man's outstretched hand. They have plump rounded bodies and short legs, but are surprisingly agile. Their feet have thick rubbery pads and sharp hooked claws, and they can scamper over steep rocks with great speed.

THE SEARCH FOR FOOD

Gundis are vegetarians and will eat the stems, flowers, leaves and seeds of just about any plant they can find. However, their teeth are not covered with the hard orange enamel that is typical of most rodents. Because of this they do not tackle woody stems or bark.

Unlike many of their rodent relatives, gundis do not store food, so a large part of each day is spent foraging. Normally the animals do not travel far from home, but when food is very scarce they may cover ½ mile in a single morning. The gundi's body is very efficient at conserving moisture. All the water the animal needs is obtained from the food it eats.

DAILY ROUTINE

Gundis leave their sleeping holes early in the morning. After checking carefully for signs of danger they feed for a while, then settle down to rest in the Sun. The Sun's heat keeps their bodies warm and helps their digestion – two good ways of making a poor food supply go a long way.

By late morning the temperature is over 86°F. The animals head for the cool shade of their holes. The evening is spent foraging again, and at dusk gundis return to their holes for the night.

CHIRPS AND CHUCKLES

Gundis are sociable creatures and live in colonies made up of many individual families. Male-female pairs and

▼Speke's gundi lives in the arid parts of East Africa. It has the largest tail of any species and uses it in social displays.

their young have small territories and "home" shelters, but they do not make permanent homes. They are easy-going animals — they often "move house" and do not usually bother chasing other gundis from their home territory.

They communicate with a variety of chirping, chuckling and whistling sounds. Some are warnings, some are greeting calls, and others help the female gundi find where she left her young hidden among the rocks.

The young are usually born in the spring. They are already covered in hair, have their eyes open and can move around soon after birth.

►**Species of gundi** Each species makes different sounds. North African gundi **(1)**, whose distinctive chirping helps members of this species recognize each other in the habitat they share with the Desert gundi (*Ctenodactylus vali*), a species that whistles. The Mzab gundi (*Massoutiera mzabi*) **(2)**, the least "talkative" gundi, has flat, immovable ears. Speke's gundi **(3)** has a rich vocabulary of sounds. The Felou gundi (*Felovia vae*) **(4)** makes a harsh *chee-chee* sound when in danger.

VISCACHAS AND CHINCHILLAS

High in the Andes mountains of Peru a scientist scans a rocky slope through binoculars. In a hollow between two boulders a small gray-brown animal is busy feeding on a patch of coarse grass. It looks like a cross between a squirrel and a large rabbit, but it is a mountain viscacha.

The viscachas and chinchillas are rodents, that is, members of the same huge animal group as rats, mice and porcupines. They inhabit some of the most remote parts of South America – the high mountains and grassland plains of Argentina, Chile, Bolivia and Peru. Yet their names are familiar to people who live thousands of miles away in New York, Paris and Tokyo. The reason is that the fur of these animals is in demand for expensive wraps and coats.

ONCE NEARLY EXTINCT

The chinchillas especially are prized for their soft, lightweight, blue-gray fur. But this sort of popularity had unfortunate results for the animals themselves. It takes over 150 pelts to make one full-length coat, and hunting for the international fur trade brought the chinchillas to the brink of extinction. At the height of this trade, 100 years ago, over 200,000 chinchilla furs a year were sold in London auction-rooms alone.

Today, chinchillas are bred on special farms to provide furs for the luxury trade. This has taken some of the pressure off the remaining wild chinchillas, but they are still rare and listed as endangered.

MOUNTAIN HIGH-LIFE

As well as providing a home for the chinchillas, the rocky slopes of the

VISCACHAS AND CHINCHILLAS

Chinchillidae (*6 species*)

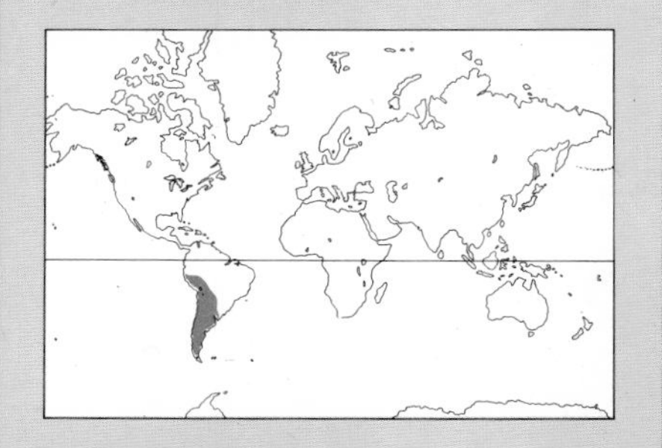

Habitat: rocky slopes at high altitude; Argentinian pampas.

Diet: plant leaves, stems and seeds; mosses and lichens.

Breeding: chinchilla: 1-6 young after pregnancy of 111 days; Plains viscacha: 1-4 young after 154 days.

Size: chinchilla: head-body 9-15in, weight 1-1½lb; Plains viscacha: head-body 19-26in, weight 9-17½lb.

Color: gray, blue-gray or brown, white or yellow-white undersides.

Lifespan: up to 10 years (chinchilla).

Species mentioned in text:
Chinchillas (*Chinchilla* species)
Plains viscacha (*Lagostomus maximus*)
Mountain viscachas (*Lagidium* species)

▼Young chinchillas, often kept as pets, are able to nibble plants less than an hour after they are born.

high Andes mountains also support three kinds of mountain viscacha. These hardy little animals can live as high as 16,000ft, feeding during the day on the stems and seeds of mountain grasses, and on the mosses and lichens that grow on the rocks.

Mountain viscachas live in colonies of up to 80 animals, using natural holes and spaces among the rocks as shelter from the wind and cold. They are very sociable animals and often cuddle up to each other when sitting out in the Sun.

They communicate with whistling calls. A long note signals that a large animal such as a human or dog or sheep is near by. Most of the animals "freeze" and wait to see what happens next. Others climb higher to get a better view. A short sharp call is the signal that a hawk has been spotted. This immediately sends every animal diving for the safety of a hole among the rocks.

Female mountain viscachas may produce young up to three times in a year, but they usually have only one baby at a time. Food is scarce in the mountains, and it is unlikely that a female could produce enough milk to feed more than one.

PAMPAS LOW-LIFE

Down on the vast plains of the Argentinian pampas lives the much bigger Plains viscacha. Males weigh up to 18lb (ten times as much as a chinchilla) and females about 9lb. The viscachas live in underground burrow systems called *viscacheros*. Each group of 15 to 30 animals is led by a senior male and may contain animals of several generations.

Unlike its mountain cousins, the Plains viscacha is most active at night. It leaves the burrow as evening falls and feeds on any vegetation it can find. A group will strip the ground over the burrow completely bare – 10 viscachas eat as much as a sheep, and their acidic urine destroys pasture – which may help them to see foxes and other predators in good time. The Plains viscacha also has the odd habit of collecting sticks, stones and bits of bone and piling them in heaps over the burrow entrances.

▼A viscacha mother and young. Females reach maturity when 2 years old, but males mature in 7 months.

MOLE-RATS AND ROOT RATS

It is midday in the Sun-baked landscape of Ethiopia. A group of biologists seek shade to avoid the blistering 140°F heat. Yet just 8in beneath their feet a "chain-gang" of small, hairless, burrowing animals are hard at work on a new feeding tunnel. They toil away in a pitch-dark world where the temperature remains at a steady 86°F. This is the world of the Naked mole-rat.

MOLE-RATS AND ROOT RATS
Bathyergidae and Rhizomyinae
(*15 species*)

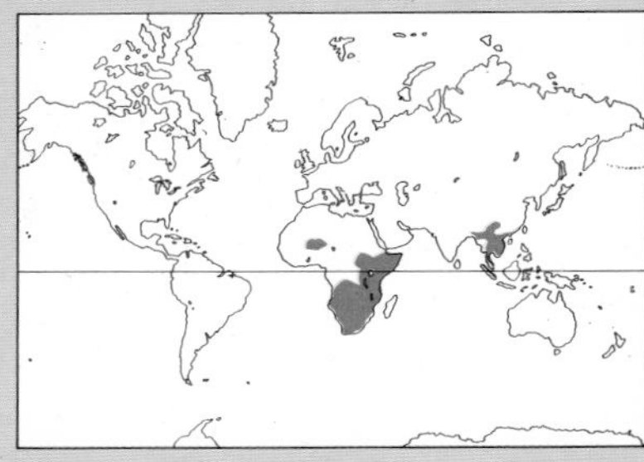

Habitat: underground burrows in sand and various different soils.

Diet: roots, tubers, some species also herbs and grass.

Breeding: litters of about 12 after pregnancy of 70 days (Naked mole-rat, details of Common mole-rat and others unknown).

Size: smallest (Naked mole-rat): head-body 3½-5in, weight 1-2 ounces; largest (Dune mole-rat): head-body to 12in, weight 1½-4lb; root rats larger.

Color: dark brown to light brown, Naked mole-rat pale pink color.

Lifespan: to 10 years in captivity.

Species mentioned in text:
Common mole-rat (*Cryptomys hottentotus*)
Dune mole-rat (*Bathyergus suillus*)
East African root rat (*Tachyoryctes splendens*)
Naked mole-rat (*Heterocephalus glaber*)

Mole-rats live across most of Africa south of the Sahara Desert. They are found in many different habitats, wherever there is soft soil or sand for them to burrow in.

TUNNELLING RATS

The mole-rats are rodents that have become completely adapted to life below ground. Their bodies are almost cylindrical in shape, to fit into their tunnels. Their eyes are tiny, and they can close their nostrils to keep out the soil as they dig.

The most unusual thing about these animals is the digging method they use. Most burrowers dig with spade-like front feet, but mole-rats dig with their enormous front teeth. As the animal chisels its way through the soil, it pulls the loose earth under its body with its front feet, then pushes it back with its hind feet.

LIFE IN THE DARK

Because they live in complete darkness, the mole-rats depend mainly on their senses of touch and smell to find food and communicate with each other. Most of them have very sensitive hairs scattered over the body to help them feel their surroundings. For the Naked mole-rat these are the animal's only hairs.

Mole-rats live on roots and on the underground tubers and bulbs that

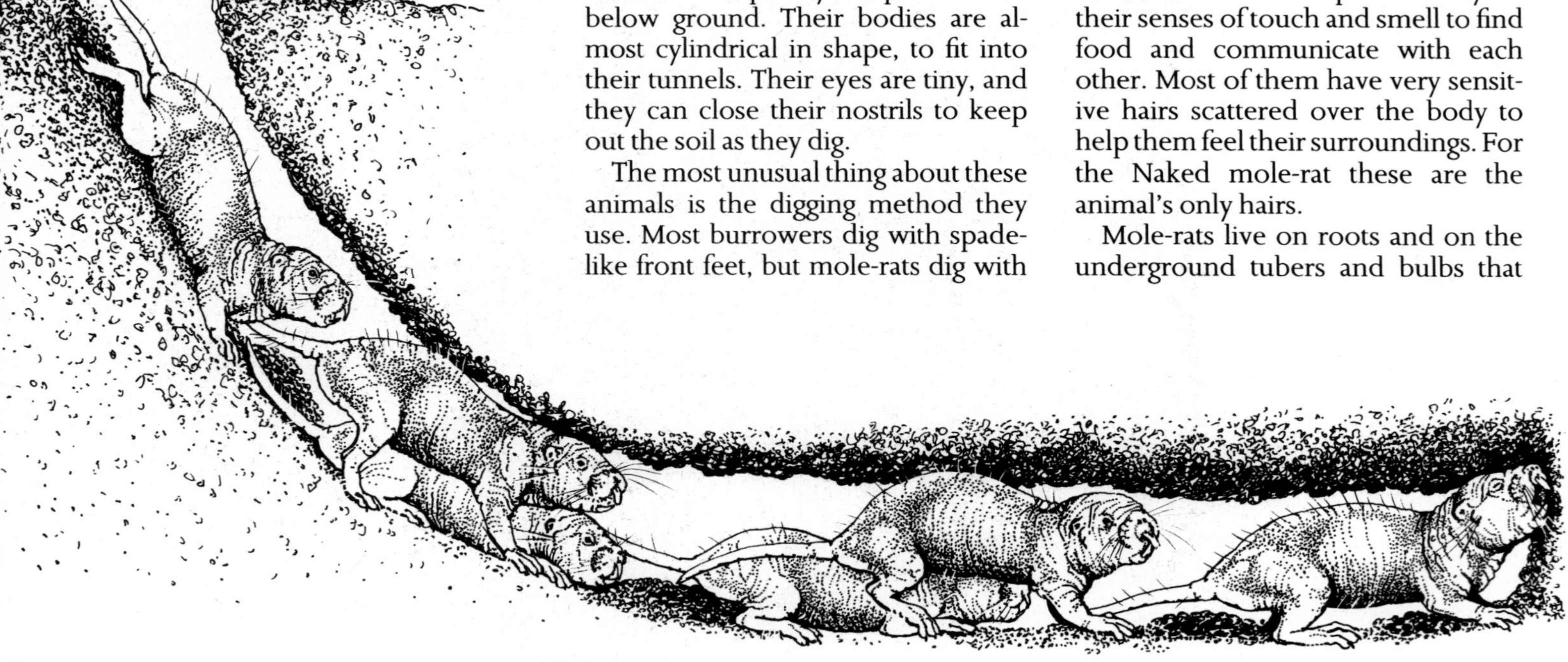

▼**The mole-rat chain gang** Naked mole-rats are organized rather like bees and ants, with different groups doing different jobs. "Workers" dig new tunnels. Each mole-rat takes a turn digging at the front, then pushes the loose earth backwards up the tunnel. The animal then works its way back to the front of the line by stepping over its team-mates. At the exit hole, a "kicker" mole-rat shoves out the earth.

plants use as food stores. They gnaw large tubers where they are and leave them to continue growing. They collect smaller ones, storing them in underground larders.

Mole-rats collect all their food in the tunnel system, which also contains resting rooms and toilet chambers. The systems can be up to ½ mile long.

UNDERGROUND TEAMWORK

The odd one out in the mole-rat group is the Naked mole-rat. Most species live alone or in small groups, but the Naked mole-rat lives in colonies of up to 80 animals. Only one pair breeds. The rest are organized as "workers," who dig tunnels and collect food, or "assistants," who look after the nest and the breeding female.

▲The young Naked mole-rat is suckled only by the mother. Later it is fed on plant food by all the colony members.

▼The East African root rat belongs to a different family from mole-rats, but it has the same life-style.

RABBITS AND HARES

In the late afternoon of a spring day, a male and female hare face each other at the edge of a field. For several minutes they glare at each other without moving. Suddenly the female hare leaps right over the male, giving him a vicious two-footed drop-kick as she goes by. The male hare spins round and kicks back. Then the two animals drop back on their haunches and sit upright, face to face, cuffing and boxing furiously with their forepaws. Eventually the male decides he has had enough. He turns to run. But even as he dives for the safety of the nearby hedge, the female victor gets in one last kick.

The "mad March hares" that perform such crazy antics during the mating season belong to one of the world's most successful animal groups. The common European rabbit is the most familiar rabbit of open fields and grasslands, but it has cousins that live in the snow-covered Arctic wilderness, and others that live in deserts and tropical forests and even on the tops of mountains.

ACROSS THE GLOBE

World wide there are 44 species of rabbit and hare. There are native species in North America and parts of South America, all over Africa and right across Europe and Asia. Some parts of South America, and all of Australia and New Zealand, originally had no native rabbits and hares, but European animals were taken there by settlers and very quickly made themselves at home. In Australia the European rabbit is still a pest. Only Antarctica and the huge island of Madagascar are completely without rabbits and hares.

RABBITS AND HARES Leporidae (*44 species*)

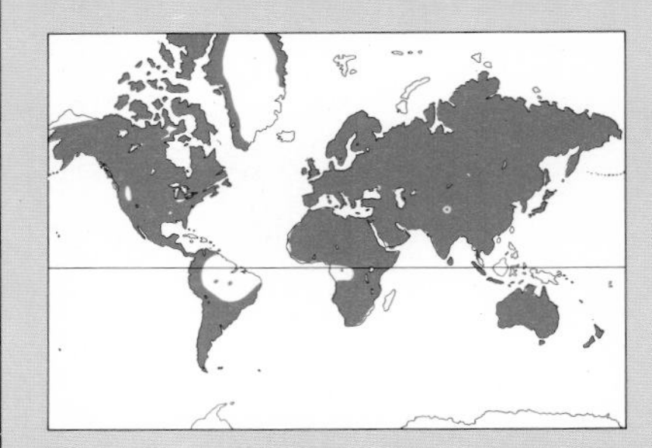

Habitat: virtually everywhere, from coast fields, grassland and forest to desert and high mountains.

Diet: grass, leaves, bark; also crops and tree seedlings.

Breeding: rabbits: litters of 3-12, up to 5 times a year, after pregnancy of 30-40 days; hares: litters of 1-9, up to 4 times a year, after pregnancy of up to 50 days; northern species mainly spring/summer; others breed throughout year.

Size: smallest (Pygmy rabbit): head-body 10-12in, weight 8 ounces; largest (European hare): head-body 20-30in, weight 11lb.

Color: reddish-brown through brown, buff and gray to white; Arctic species change from brown to white in winter.

Lifespan: usually less than 1 year; domestic (pet) rabbits up to 18 years.

Species mentioned in text:
Arctic hare (*Lepus timidus*)
Black-tailed jackrabbit (*L. californicus*)
Bushman hare (*Bunolagus monticularis*)
Cottontails (*Sylvilagus* species)
European hare (*Lepus europaeus*)
European rabbit (*Oryctolagus cuniculus*)
Hispid hare (*Caprolagus hispidus*)
Pygmy rabbit (*Sylvilagus idahoensis*)
Snowshoe hare (*Lepus americanus*)
Sumatran hare (*Nesolagus netscheri*)
Volcano rabbit (*Romerolagus diazi*)

► Rabbits and hares of the world The Hispid hare **(1)** is an endangered species from India. It seldom leaves the shelter of the forest. Volcano rabbit of Mexico **(2)** shown sitting in the long grass recycling its droppings. The European hare in its "boxing" position **(3)**. Greater red rock-hare (*Pronolagus crassicaudatus*) **(4)** of Southern Africa in alert posture. Male Eastern cottontail (*Sylvilagus floridanus*) **(5)** in alert posture.

10

9

◀The very rare Sumatran hare **(6)** grooming its muzzle and spreading scent from scent glands. Bunyoro rabbit (*Poelagus marjorita*)**(7)** hopping along. The species is common in parts of Central and Eastern Africa. Adult male European rabbbit **(8)** scratching his chin. Bushman hare **(9)** of the Southern African river banks, now an endangered species. The Amami rabbit (*Pentalagus furnessi*) **(10)** is found on just two small Japanese islands. The total population is about 5,000, and the species has protected status.

FLEET OF FOOT

Rabbits and hares are very similar in shape, though they vary in size. They have long soft fur, which covers the whole body including the feet, and a small furry tail which is usually white, or at least white underneath. The tail is always turned upwards, so that as the animal runs, the white fur may act as a target for predators, keeping them away from vital areas of its body. The animal's alarm signal is to thump its back feet on the ground.

The front legs are quite short but strong, and the five toes on each forefoot have sharp claws for digging. The back legs are very much longer and are clearly designed for running and bounding over open ground. Some of the largest hares can reach speeds up to 50mph when fleeing from danger.

Both rabbits and hares have large eyes, placed high on the sides of the head. They give clear vision in twilight and at night, which is when many species are most active. The position of the eyes also gives the animals very good vision to the sides, and even above and behind them. This is very important to an animal that makes such a tasty and tempting target for eagles, wild cats, polecats and a host of other predators.

FOOD PROCESSING

Rabbits and hares live entirely on plant food, mainly grasses, leaves, bark and roots. Like the rodents, they have two large front teeth which grow continuously. But they differ from rats and mice by having a second, much smaller pair of front teeth tucked in behind the main pair.

Their internal organs too are especially developed to cope with large amounts of low-quality vegetable food. The ground-up food passes into the stomach, and then into the gut. But instead of passing straight out through the last section of the gut it is held for a while in another stomach-like bag where bacteria help to break down the coarse food more thoroughly.

Rabbits and hares produce two kinds of dropping. When the animal is most active, normal firm dry droppings are left, often in special "latrine" areas. But when the animal is resting, much softer droppings are produced. These are eaten again and recycled through the digestive system for a second time so that useful chemicals such as vitamin B can be absorbed into the animal's body.

Because rabbits breed very quickly, and can eat a great variety of food, they can easily become a nuisance. In some areas they are "public enemy number one" for farmers and foresters. In the United States, for example, Black-tailed jackrabbits cause widespread damage to crops in California, while the cottontails and the Snowshoe hare can ruin new forestry plantations by nipping the growing shoots off the tender young tree seedlings.

BURROWS AND HOLLOWS

The biggest differences between rabbits and hares can be seen in their choice of where to live and in the way they bring up their young.

Most rabbits live in underground burrow systems called warrens. The young rabbits (kittens) are born in a warm nest, snugly lined with hair and soft grass, either in the main warren or in a nursery burrow near by. They are hairless, and their eyes do not open for several days (10 days for European rabbits). The female (doe) feeds her young for only a few minutes in each 24 hours. She then seals them in by covering the burrow entrance with earth, and goes off to feed.

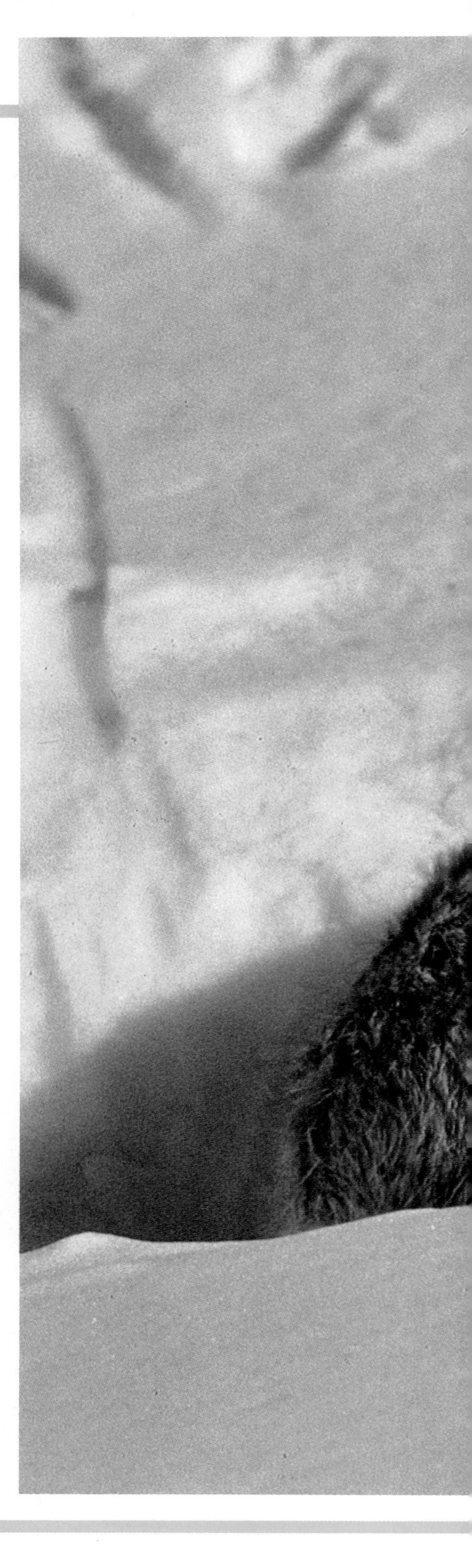

► Many northern species, like this jackrabbit, shed their fur in the fall and replace their light brown summer coat with a much thicker, warmer, winter coat of white.

►Outside the breeding season, the European hare is a shy, quiet animal. But all that changes in the spring. As the mating instinct takes over, the males (bucks) and females (does) become aggressive and quite outrageous in their behavior. They dash about, leaping into the air, fighting and chasing each other. As each male tries to mate with a female that is ready to breed, she rejects the advances of any male she considers unworthy.

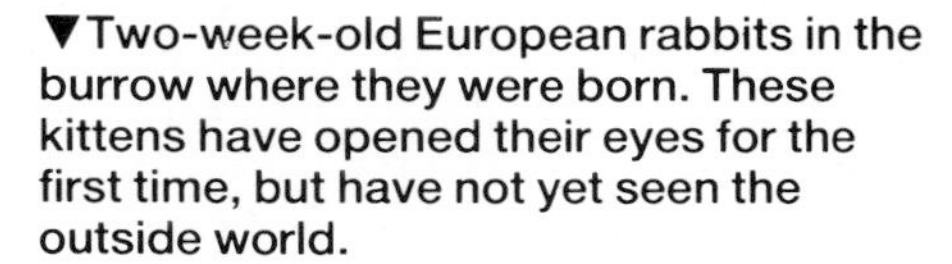

▼Two-week-old European rabbits in the burrow where they were born. These kittens have opened their eyes for the first time, but have not yet seen the outside world.

The young rabbits do not venture outside for about 3 weeks.

Hares are very different. Only a few of them make burrows of any kind. Usually they rest in a shallow hollow, called a form, in soft earth or in long grass. The young hares (leverets) are well developed at birth. They are covered with warm fur, and their eyes are open. When they are just 2-3 days old the mother places each one in a separate form, well hidden among rocks or tall grass. There they remain until the family meets up, usually around sunset, for the one feed of the day. Because they live out in the open, young hares can run almost from birth. Their main protection, however, is to remain absolutely still.

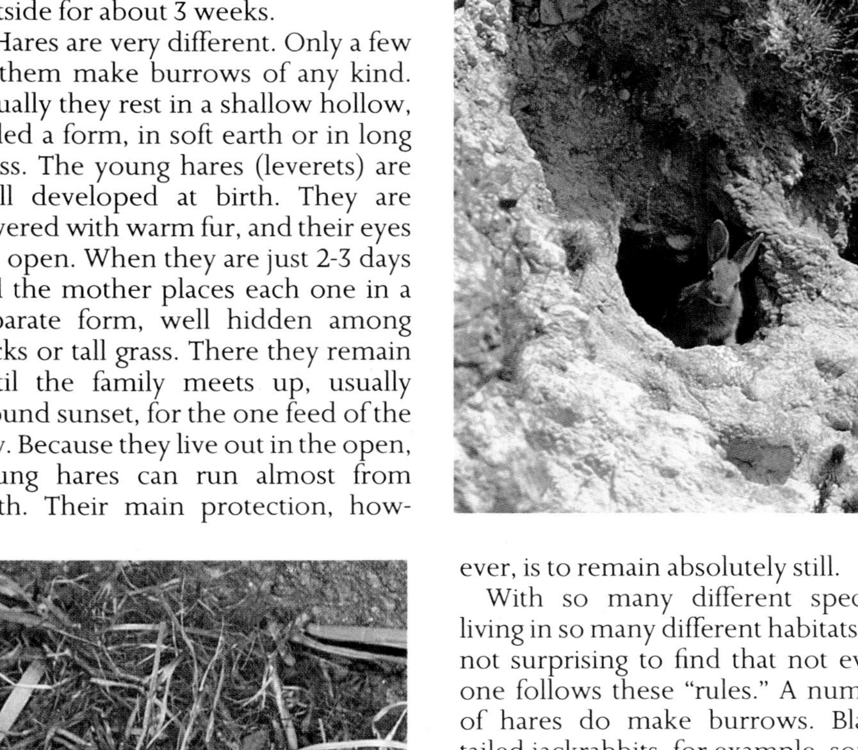

◄In sandy soils, rabbits can easily dig new burrows. But in areas of hard soil, like this chalky bank, the best burrow sites are prized possessions.

With so many different species, living in so many different habitats it is not surprising to find that not every one follows these "rules." A number of hares do make burrows. Black-tailed jackrabbits, for example, sometimes dig short burrows to escape from the fierce summer heat in the American deserts. Arctic hares in Scotland may dig burrows for their young to use in times of danger. Not all rabbits dig burrows either. Many of the cottontails either use holes made by other animals or simply hide themselves among thick vegetation.

THE NUMBERS EXPLOSION

European rabbits are famous for the speed at which they reproduce. Females often produce five litters in a year, each of 5 or 6 young (occasionally up to 12). Each young female, in turn, will be ready to produce her own first family by the time she is 3 months old. At that rate it is no surprise that rabbits can quickly become a major pest.

Hares do not multiply quite so quickly, but they too are fast breeders.

The European hare was taken to Argentina in 1888. In just 100 years it has spread throughout the whole country. It is now found spread over 2 million sq miles. In the central pampas region, 5-10 million hares are caught each year for their light meat.

RARE RABBIT RELATIVES

The family that contains one of the most common animals on Earth also includes several surprisingly rare species. Some are on the international list of endangered species.

One of the rarest is the Sumatran hare, which inhabits the remote mountain forests of the South-east Asian island of Sumatra. Only 20 have ever been seen, and only one has been seen in the last 10 years. If the forests are cut for their valuable timber, this unusual striped hare will disappear for ever.

In the sal forests of northern India and Bangladesh, the Hispid hare is also becoming more and more rare.Its woodland habitat is being destroyed to make way for cattle grazing. The Bushman hare of Southern Africa faces a similar threat. Its natural habitat is the dense vegetation along river banks, but these are also the most fertile areas and so they are being taken over by farmers, leaving the hares with nowhere to go.

Strangest of all is the case of the Volcano rabbit, which lives at 10,000-13,000ft on the flanks of two volcanic mountain ranges near Mexico City. It is one of the world's smallest rabbits, and lives in groups of up to five animals in burrows among open pine-woods and grassy slopes. It is active mainly during the day, using a variety of calls to keep in touch with one another. Unfortunately the 17 million people of Mexico City are barely half an hour's drive away, so the rabbits are threatened now by hunters and noisy tourists as well as by destruction of their habitat.

PIKAS

High on a scree slope in the Rocky Mountains of Colorado, a large male pika lets out a high-pitched squeak of rage. Leaping from his favorite rock he charges at a young male who has strayed into his territory. The youngster is no match for him and turns and flees with a squeal of alarm.

Such fierce territorial behavior is typical of pikas. They are small furry animals, closely related to rabbits and hares, and they live in some of the toughest and most desolate habitats on Earth. There are 14 species altogether. Two are natives of America's Rocky Mountain range. The rest are spread across Europe and Asia, from Afghanistan through the Himalayas to China, Japan and Siberia.

Most species, including the Collared pika of Alaska and Canada and the Large-eared pika of Nepal, make their homes among the scree and boulders on high mountain slopes. Others, like the Steppe pika, inhabit the dry grasslands and semi-desert regions of central Asia, and make their homes in burrows.

Throughout their range, no two species share the same habitat. If two species live near each other, one will take the rocky areas, the other will take the grassland.

PIKAS Ochotonidae (*14 species*)

Habitat: rocky mountain slopes to over 20,000ft, steppe and semi-desert.

Diet: grass, leaves, flowers and bark.

Breeding: average litter 5 in rock-dwellers, larger in burrowing species. Pregnancy 25-31 days.

Size: head-body 7-8in; weight 2-7 ounces in Steppe pika to 6½-10 ounces in Afghan.

Color: pale yellow-brown to darker reddish-brown. Paler underneath.

Lifespan: to 7 years.

Species mentioned in text:
Afghan pika (*Ochotona rufescens*)
Collared pika (*O. collaris*)
Large-eared pika (*O. macrotis*)
North American pika (*O. princeps*)
Steppe pika (*O. pusilla*)

▼During the breeding season, male pikas give long "singing" calls lasting up to half a minute. Short squeaks are used to show anger or fear, or to warn others that a predator is in the area.

▲The Collared pika inhabits rocky slopes in north-western Canada and Alaska. Like the North American pika, it came to the American continent from Asia when the two land areas were still joined together some 10,000 years ago.

◀Hay gathering gets more and more frantic towards the end of summer. In some species, hay stores are jointly owned by pairs of males and females. In others they are strictly private property.

THE HAY-MAKERS

One of the most unusual things about pikas is that all but the two Himalayan species are hay-makers. Throughout the summer and the fall the animals spend a large part of their time gathering grass, leaves, flowers and moss which they store for winter food. Rock-dwelling pikas, often pile their hay beneath overhanging rocks. Burrow-dwellers make small haystacks near their homes, either in the open or beneath a nearby bush. The haystacks can weigh 30 to 40 times as much as the animals that built them.

The hay store is an important source of food when the land is covered in snow, but the pikas also tunnel under the snow to gnaw the bark of apple trees, aspens and young conifers.

Like rabbits and hares, pikas produce two kinds of droppings. Daytime droppings are dry, rounded pellets which are left on the ground. Nighttime droppings are soft and dark green and are rich in energy and certain vitamins. The pikas eat these so that none of the food value goes to waste.

WHAT SIZE OF FAMILY?

The size of pika families depends on where they live. Typical rock-dwellers have litters of 5 or 6 young, but the burrow-dwellers have larger families. The young of the rock-dwellers are also much slower to develop. They do not breed until they are at least a year old, while the young of the burrowers usually breed in their first summer.

Pikas have a very complex social system. Each animal has its own small territory, which it defends jealously, especially against intruders of the same sex. To keep down the number of fights, males' territories are usually separated from each other by those of females. Young male pikas must wait until they can take over a territory by force or one becomes free when an adult male dies or is killed by a hawk.

COLUGOS

As dusk falls in a Malayan coconut plantation a colugo leaves the tree-hole in which it spent the day asleep. Gripping the trunk with needle-sharp claws, it climbs to the top, then launches itself into space. It glides through the trees to land on a trunk over 330ft away.

COLUGOS Cynocephalidae
(*2 species*)

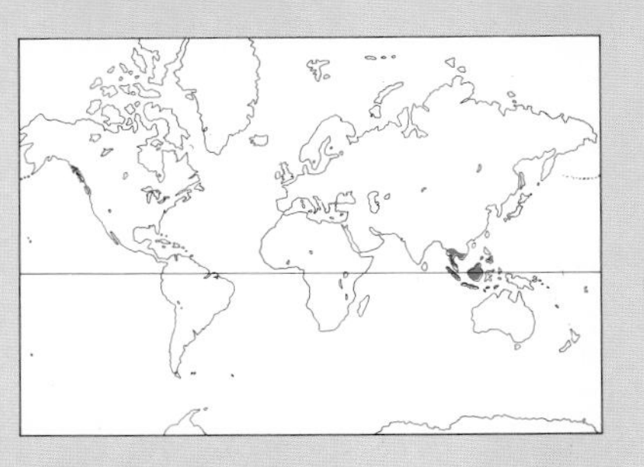

Habitat: tropical rain forest, rubber and coconut plantations.

Diet: leaves, shoots, buds and flowers; some soft fruits.

Breeding: 1 young (rarely 2) after pregnancy of 60 days.

Size: Malayan colugo: head-body 13-17in, tail 9-11in, "wingspan" 28in, weight 2-4lb. Philippine colugo: head-body 13-15in, tail 9-11in, weight 2-3½lb.

Color: back, mottled gray-brown with white spots; paler underneath. Philippine species darker.

Lifespan: not known.

Species mentioned in text:
Malayan colugo (*Cynocephalus variegatus*)
Philippine colugo (*C. volans*)

The two species of colugo of Southeast Asia puzzled scientists for many years. Some thought they were lemurs – the animals are often called "flying lemurs" – others grouped them with the bats or the insect-eaters. Now we know they are none of these things. They are the only survivors of an ancient and very specialized group of animals, and they even have their own group name Dermoptera, which means skin-wing.

NATURE'S HANG-GLIDERS

It is dark inside a tropical rain forest. Tall straight tree trunks tower 100ft or more above the ground. Most of the food is up there too, high in the leafy canopy. Birds and bats get to the food by flying. Squirrels, cats and monkeys reach it by climbing. But a colugo scrambles up there then glides from tree to tree.

Tough, flexible flaps of skin stretch from the sides of a colugo's neck to the tips of its fingers and toes, and to the end of its tail. At rest, the skin hangs around the animal like a loose cloak, but when it leaps from a tree and stretches out its legs, the skin is pulled tight into a perfect kite shape.

With slight movements of its body a colugo can steer its chosen path through the air. One glide measured by scientists carried a colugo 150yd to a landing point only 13yd lower than its take-off point.

FOREST HIGH-LIFE

Colugos are so specialized for life in the trees that they can hardly move if placed on the ground. They climb with the aid of sharp claws, reaching up to grip a tree trunk with their front paws, then bringing up both back feet together.

Both species spend the daylight hours at rest in a tree-hole or clinging to a tree trunk. In plantations they may curl up in the middle of a palm frond. At dusk they emerge to feed on the forest vegetation, pulling small branches within reach, then stripping off the leaves with their strong tongue and sharp teeth. Colugos move through the forest as they feed, and will use favorite gliding trees again and again. Where the feeding areas of several colugos overlap, they often share particularly good take-off trees.

►The dappled gray and brown coloring of the Malayan colugo provides a very effective camouflage against tree bark.

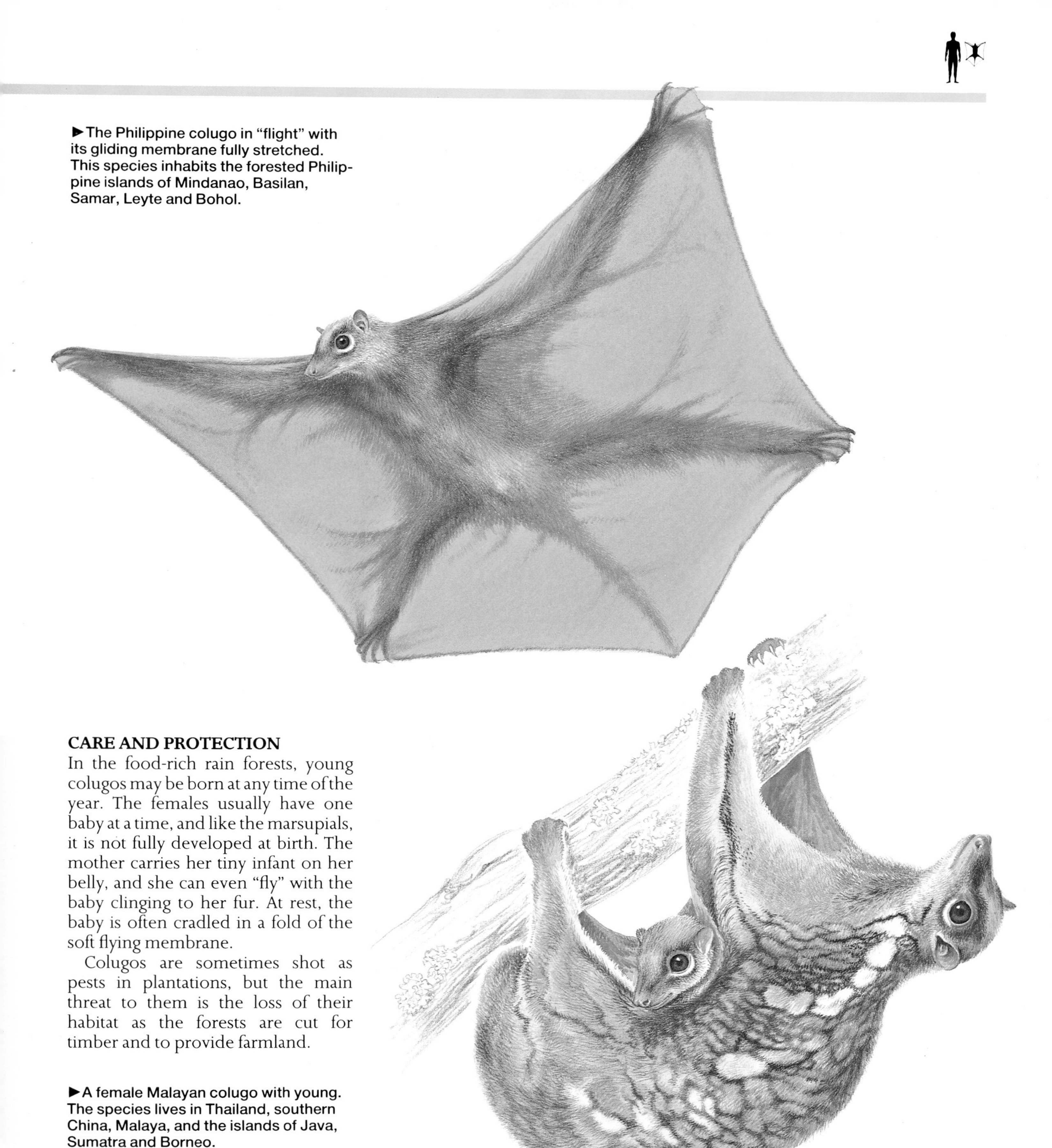

►The Philippine colugo in "flight" with its gliding membrane fully stretched. This species inhabits the forested Philippine islands of Mindanao, Basilan, Samar, Leyte and Bohol.

CARE AND PROTECTION

In the food-rich rain forests, young colugos may be born at any time of the year. The females usually have one baby at a time, and like the marsupials, it is not fully developed at birth. The mother carries her tiny infant on her belly, and she can even "fly" with the baby clinging to her fur. At rest, the baby is often cradled in a fold of the soft flying membrane.

Colugos are sometimes shot as pests in plantations, but the main threat to them is the loss of their habitat as the forests are cut for timber and to provide farmland.

►A female Malayan colugo with young. The species lives in Thailand, southern China, Malaya, and the islands of Java, Sumatra and Borneo.

HYRAXES

On top of a rocky outcrop in Tanzania's Serengeti National Park, about 30 hyraxes huddle together in the early morning Sun. As they warm up, the youngsters begin to play, and some of the adults wander off to feed. Suddenly the peace is broken by a shrill whistle of alarm from a large male acting as look-out. The hyraxes dive for cover as an eagle plunges out of the Sun.

Anyone seeing a hyrax for the first time would probably think it was a relative of the guinea pigs or the hamsters. But far from being related to the rodents, the hyrax is a distant relative of the biggest land animal on Earth – the elephant! The connection is a very distant one, but both elephants and hyraxes are descendants of the same group of prehistoric hoofed animals.

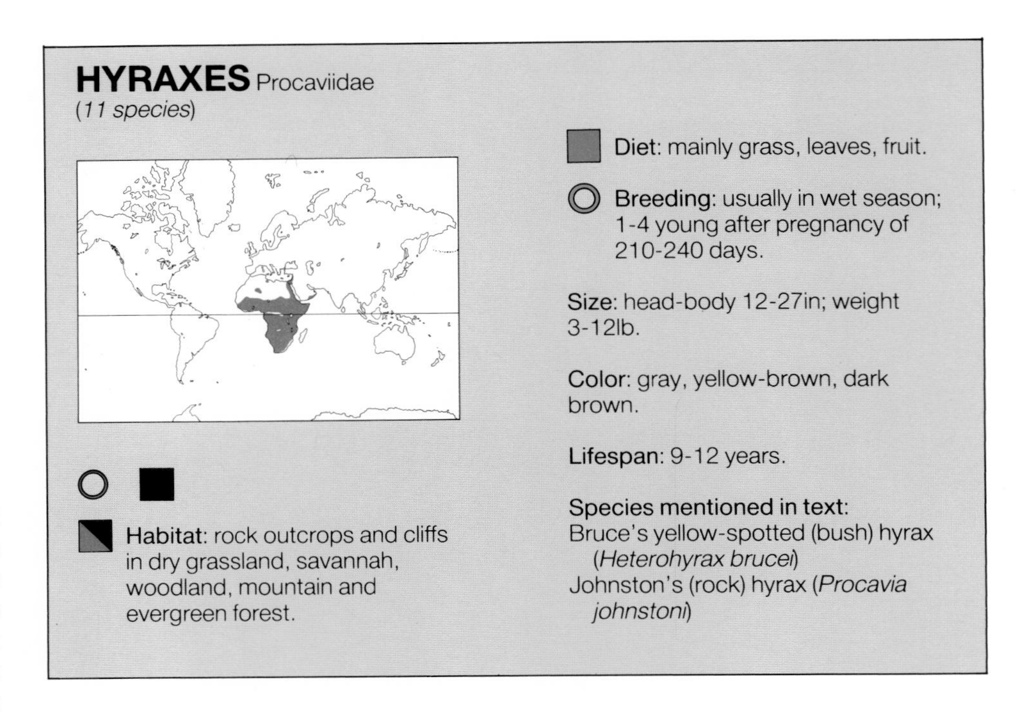

HYRAXES Procaviidae
(*11 species*)

Habitat: rock outcrops and cliffs in dry grassland, savannah, woodland, mountain and evergreen forest.

Diet: mainly grass, leaves, fruit.

Breeding: usually in wet season; 1-4 young after pregnancy of 210-240 days.

Size: head-body 12-27in; weight 3-12lb.

Color: gray, yellow-brown, dark brown.

Lifespan: 9-12 years.

Species mentioned in text:
Bruce's yellow-spotted (bush) hyrax (*Heterohyrax brucei*)
Johnston's (rock) hyrax (*Procavia johnstoni*)

▼Hyraxes, like this Johnston's hyrax of north-east Africa and the south-east Arabian Peninsula, will put up a ferocious fight if attacked or cornered.

►Bush hyraxes, such as Bruce's yellow-spotted hyrax, sometimes inhabit hollow trees, but usually they live in holes in rock outcrops.

The 11 present-day species of hyrax are found throughout Africa south of the Sahara, and one group, the rock hyraxes, also live in the Middle East.

RUBBER-SOLED SPRINTERS

Despite their chunky bodies and short legs, hyraxes are outstanding climbers and jumpers. The undersides of their feet have thick rubbery pads with large numbers of sweat glands, and as the animals run the sweat makes the pads sticky. This allows the hyraxes to dash up and down steep rocks and tree trunks with astonishing agility. This speed is important to the hyraxes' survival, for these small mammals are favorite prey for eagles, owls, leopards, jackals, hyenas and many kinds of snake.

KINGS OF THE KOPJES

Hyraxes are divided into three groups. Most widespread are the rock hyraxes or dassies of the savannah woodlands, plains and highlands. Like the bush hyraxes of East Africa, they live on small rocky outcrops locally called kopjes, and on cliffs. Tree hyraxes inhabit woodland and rain forests across central Africa, but they too are found among the barren rocks of the Ruwenzori mountains.

All the hyraxes are plant-eaters. Rock hyraxes eat mainly coarse grass while the others eat softer leaves, fruit and occasionally birds' eggs and insects. Strangely, a hyrax's teeth are not well suited to this diet. The tusk-like incisors are hardly used so the animal has to turn its head to one side and use its sharp cheek teeth.

Most of the hyraxes live in dry habitats so their bodies are designed to conserve water. Their urine is very concentrated, and contains large amounts of waste chemicals, including calcium carbonate. As the animals always use the same toilet areas (latrines), the rocks there soon become coated with white limestone crystals.

LIVING TOGETHER

Hyraxes are very sociable animals. They live in groups of up to 50, and a group will often contain roughly equal numbers of two different species.

Living in groups has many advantages. Hyraxes are not very good at controlling their body temperature, so when they leave their sleeping holes in the morning they huddle together for warmth. There is safety in numbers too. Whenever the animals are feeding, at least one senior adult is always on guard duty.

Even though two species may live together, each will breed only with its own kind. Females become pregnant once a year, and young are usually born in the rainy seasons, when food is in plentiful supply.

SLOTHS

Tree-fellers in a Central American rain forest stop their destruction to watch a strange animal move through the nearby tree canopy. It is a sloth. The animal can barely walk, and mostly hangs upside down in trees. It travels no more than about 40 yd a day going at full speed. But in the forest it can find a supply of food all year round.

The five species of sloth are split into two families according to the number of claws on their front feet. As the name suggests, the three-toed sloths have three claws, while the two-toed sloths have just two. They all have three claws on their back paws.

These strong, curved, 4in-long hooks help the sloth climb easily among the branches. But on the ground the animal is almost helpless and can only manage a clumsy crawl. Surprisingly, sloths are very good swimmers, although they rarely leave the safety of the trees.

PORTABLE CAMOUFLAGE

The sloth's long, thick, shaggy hair is grayish-brown in color, but the animal usually has a greenish tinge. The color comes from algae (microscopic plants) that grow in tiny grooves on each hair. This portable plant-life helps camouflage the sloth as it moves among the branches. Because of its upside-down life-style, the sloth's hair grows in the opposite direction to that of most animals. It grows from the belly towards the back so that rainwater runs off easily.

The sloth's coloring, slow motion and very unusual position hanging beneath the branches all help keep it safe, even in a habitat full of fierce predators.

WHO NEEDS ENERGY?

Sloths live entirely on leaves, which they grind to a pulp with their large cheek teeth. They have no front teeth. It is a low-energy diet, but the sloth needs little fuel. It uses energy at only half the rate of most mammals of similar size.

To conserve energy it lets its body temperature drop during the cooler

SLOTHS Megalonychidae and Bradypodidae (*5 species*)

Habitat: mainly lowland and upland rain forest.

Diet: tree leaves, fruit.

Breeding: 1 young after pregnancy of 6 months, 11½ months for Hoffmann's sloth.

Size: three-toed sloths: head-body to 24in, tail 2½-3in, weight 8-10lb; two-toed sloths: head-body to 28in, no tail, weight 9-18lb.

Color: gray-brown to beige, with green tinge due to algae on hair.

Lifespan: up to 12 years, 30+ in captivity.

Species mentioned in text:
Brown-throated three-toed sloth (*Bradypus variegatus*)
Hoffmann's two-toed sloth (*Choloepus hoffmanni*)
Pale-throated three-toed sloth (*Bradypus tridactylus*)

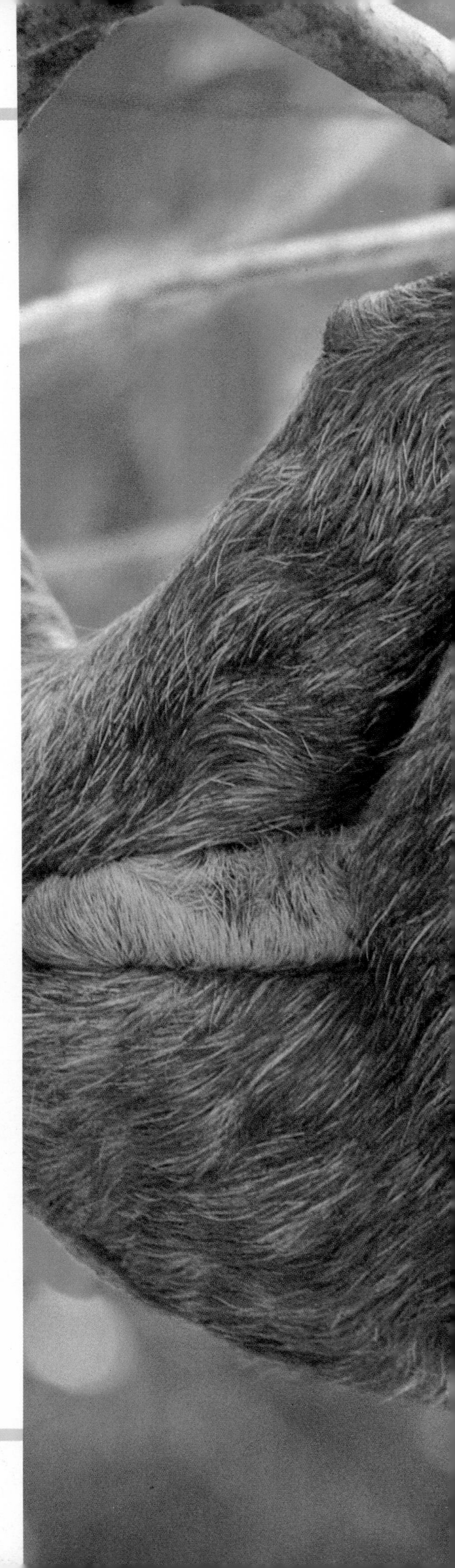

◀After 6-9 months this young Brown-throated three-toed sloth will inherit part of its mother's feeding area and also her preference for the leaves of certain tree species.

▼Hoffmann's two-toed sloth in the rain forest of Panama.

hours of night and when resting. To warm up again it simply moves out into a patch of strong sunlight.

LIVING HAMMOCK

Adult sloths are solitary for most of the year. Breeding males advertise their presence by marking tree branches with a powerful scent produced by their anal glands.

Most sloths breed at any time of the year. The exception seems to be the Pale-throated three-toed sloth of Guyana, which breeds only after the rainy season. A single baby is born in all species and is cradled on the mother's belly, where it will remain for up to 9 months. It feeds on milk for only about 1 month and from then on feeds on the leaves it can reach from its mobile hammock.

PHALANGERS

A gardener living in the suburbs of an Australian city wakes one morning to find the buds stripped from her best rose bush. The night-time visitor that ate the buds is a Common brushtail possum. The possum devours the leaves, buds and shoots of other plants too.

PHALANGERS
Phalangeridae (*14 species*)

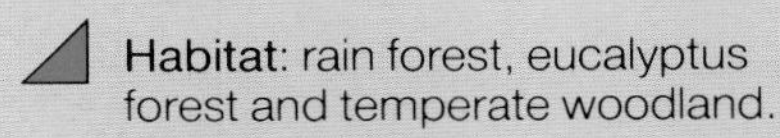

Habitat: rain forest, eucalyptus forest and temperate woodland.

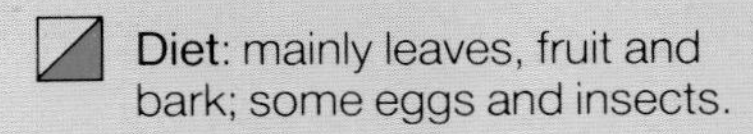

Diet: mainly leaves, fruit and bark; some eggs and insects.

Breeding: 1 or 2 young a year, in fall/spring. Pregnancy 17 days.

Size: head-body 13½-28in, tail 12-20in, weight up to 11lb.

Color: silver, gray, black or reddish brown. Some species spotted or striped.

Lifespan: to 13 years, 17 in captivity.

Species mentioned in text:
Common brushtail possum (*Trichosurus vulpecula*)
Mountain brushtail possum (*T. caninus*)
Scaly-tailed possum (*Wyulda squamicaudata*)
Spotted cuscus (*Phalanger maculatus*)
Woodlark Is. cuscus (*P. lullulae*)

The Common brushtail possum is the most widespread of the 14 species that make up the phalanger family.

IN THE TREES AFTER DARK

All the phalangers are nocturnal creatures, and all have the large eyes typical of animals that are out and about after dark. Most of them are also arboreal, that is, they spend most of their time in the trees. Their feet and tails are specially adapted for this life-style. The toes of their front feet are equipped with strong hooked claws, ideal for climbing and for pulling tender leaves within reach. However, their back feet are quite different. They have no claws, but the first toe can be turned towards the palm, like a thumb, to provide a firm grip on slender branches. Phalangers can also hang on with their tails. To provide a good grip the end section of the tail is usually bare, and even the furry tail of the Common brushtail possum has a bare patch underneath to help improve its grip.

A WIDESPREAD FAMILY

Phalangers are found throughout the woodlands and forests of Tasmania and Australia, from dry savannah

woodlands to the wet steamy rain forests of northern Queensland. They also inhabit New Guinea, and many of the neighboring islands, as far west as Sulawesi and as far east as the Solomon Islands. Phalangers are not native to New Zealand, but the Common brushtail has been taken there and released into the wild.

BRUSHTAILS AND SCALY-TAILS

The Common brushtail possum lives in wooded habitats, farmland and city gardens all over Australia and Tasmania. It feeds mainly on eucalyptus leaves, but it also comes down to the ground to vary its diet with leaves, buds and shoots of other plants. In some areas it is a serious pest on farms and in gardens. Its close relative the Mountain brushtail possum causes great damage to pine plantations in south-eastern Australia, and often raids banana crops in Queensland.

Females of the Common brushtail usually have one baby in the fall, and many produce a second one in the spring. Mountain brushtail females never have more than one baby in a year. The baby is born in the fall, and stays in the pouch for 8 months – 2 months longer than a

▲Most Common brushtail possums have silver-gray fur, but the ones that live in Queensland are a coppery red.

◄▼Australia's possums and cuscuses The Common brushtail possum **(1)** is one of three brushtail species. All have large ears, pointed faces and furry tails. The rare Scaly-tailed possum **(2)** was only discovered in 1917. Cuscuses like this female Spotted cuscus **(3)** have rounder faces than the brushtails. Despite their name they may have plain silver fur. The closely related Gray cuscus (*Phalanger orientalis*) **(4)** lives in Queensland and New Guinea.

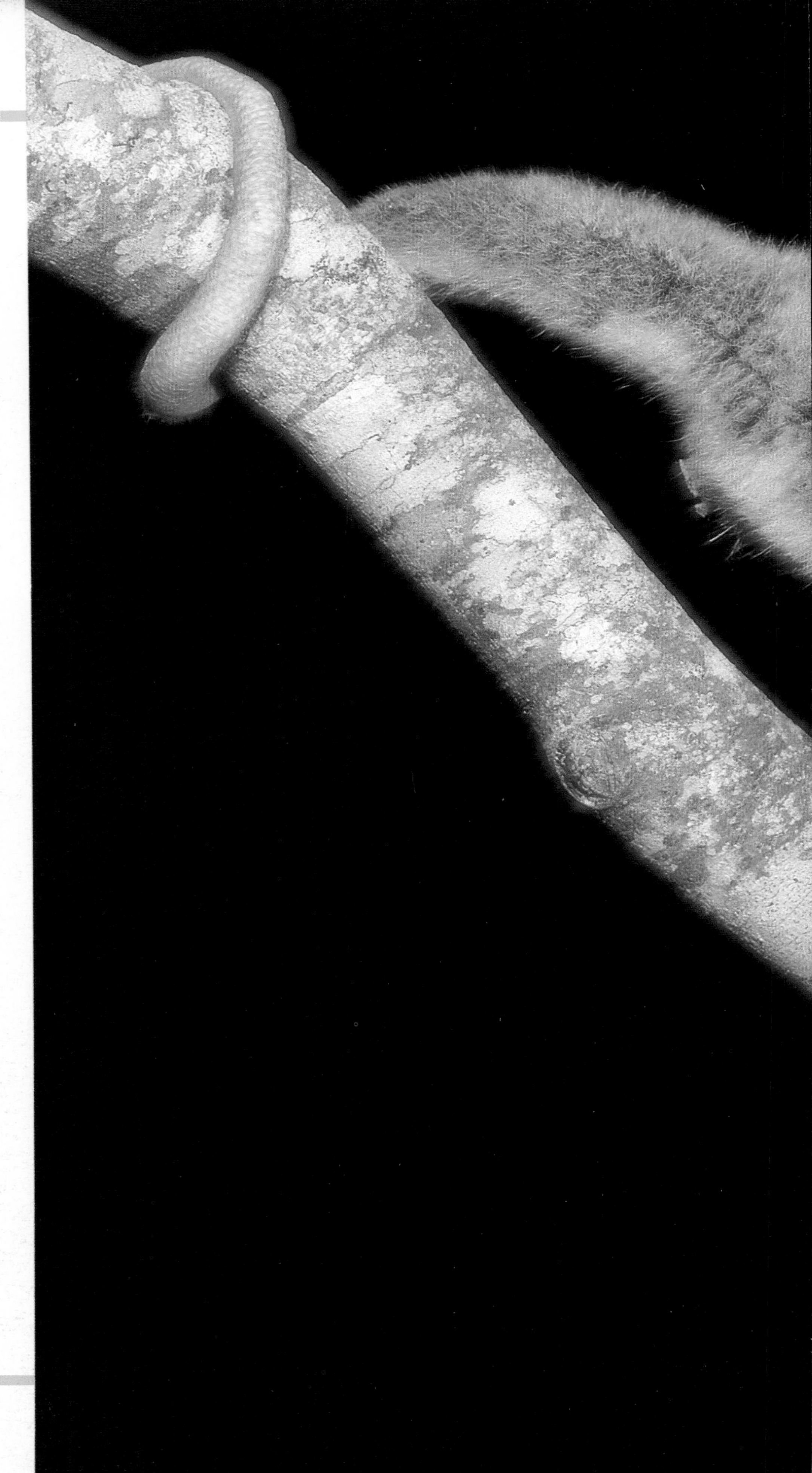

►The Spotted cuscus is common on the island of New Guinea, but is rare in the forests of northern Queensland.

Common brushtail baby stays there.

In contrast to the brushtails, the Scaly-tailed possum is quite rare. It lives in a few small areas of eucalyptus woodland near Kimberley in western Australia, feeding only on the leaves and flowers of the trees. The brushtails prefer to nest in tree-holes (dens) or even under house roofs, but the Scaly-tailed possum is thought to make its nest among rocks on the ground.

RAIN FOREST RELATIVES

The biggest group in the phalanger family consists of the cuscuses or true phalangers. There are 10 species, and they live mainly in the tropical rain forests of New Guinea and neighboring islands. On New Guinea itself, several cuscus species are found only in certain bands of forest, at different heights above sea level. These species are very similar in size and lifestyle, so by keeping to their own parts of the forest they avoid competing for food. Other cuscuses do share the same area of forest, but when this happens the larger species usually spends its time in the trees, feeding on leaves, while the smaller one lives lower down in the undergrowth, feeding on leaves, fruits, and in some cases, insects too.

For most of the year cuscuses are solitary. They mark their territories with urine and with scents from special glands, and are often very aggressive towards intruders.

None of the 14 phalangers are seriously endangered at the moment, but 3 species are already quite rare. (Only 8 Woodlark Island cuscuses have ever been seen!) In the future, the biggest threat to these forest animals will come from timber companies, cutting down the rain forests for their valuable hardwoods.

KOALA

From October to February the eucalyptus forests of eastern Australia echo with strange night-time calls. Long, harsh, indrawn breaths are followed by bellowing growls. No sooner has one call died away than it is answered by others from different parts of the forest. It is the koala's breeding season, and these are the cries of the male koalas.

KOALA *Phascolarctos cinereus*

Habitat: eucalyptus forest up to 2,000ft above sea level.

Diet: eucalyptus leaves, from a small number of preferred species.

Breeding: 1 young, in summer, after pregnancy of 34-36 days.

Size: head-body to 34in male, 30in female; weight 26lb male; 18lb female. (Animals in northern part of range are considerably smaller.)

Color: gray to reddish-brown, white on chin, chest, and under forearms.

Lifespan: 13 years, 18 in captivity.

The koala might look like a teddy bear, but it is not related to the bears at all, and is certainly not as friendly as it looks. It will defend itself fiercely with its sharp claws if attacked.

THE COMPLETE SPECIALIST

The koala is a marsupial, a pouched mammal, and one of the most specialized animals in Australia. It lives only in the eucalyptus forests of the east coast, and it eats hardly anything but eucalyptus leaves. Not only that, but with 350 eucalyptus species to choose from, the koala feeds mainly on just 5 or 6. It is a low-quality diet, not very rich in energy, so the koala is not very active. It spends almost its entire life in the trees, sleeping for up to 18 hours a day and spending the rest of its time eating. Young leaves are bitten off, then ground to a paste with the large cheek teeth. An adult koala weighing about 20lb will munch its way through up to 2lb of leaves in a day. The animal's intestine is long, to help it digest this mass of leaves.

▼Even the koala's liver is special. It deals with the poisonous chemicals that occur in some eucalyptus leaves it eats.

RAISING A BABY

For most of the year koalas live alone. Their feeding areas may overlap, but the animals do not mix very much, even when there are several of them in a small area of forest. In the breeding season, each breeding male (usually over 4 years old) has several mates. His territory overlaps theirs, and throughout the mating season he is on the move, visiting his mates, calling and bellowing, and chasing rival males from his territory.

In midsummer the female produces a single baby, and like all marsupials it crawls straight into the pouch. There it stays for about 6 months until it is fully developed and ready to cope with the outside world.

The weaning process that follows is very unusual. At first, the baby koala is fed on partly digested leaf pulp that has already passed through the mother's body. This processed food is easy to digest, but it has another important function. Along with the pulp, the young koala receives a supply of microbes from the mother's gut. These are helpful "bugs" that remain in the youngster's body and enable it to digest tough eucalyptus leaves for itself.

►▼Once it has left the pouch, a young koala rides about on its mother's back for another 4 to 5 months. The adult is a good climber. It can grasp thin branches with an unusual grip – two fingers at one side of the branch, three at the other.

WOMBATS

On a small wooded island between Australia and Tasmania a scientist settles down to study the creatures of the night. For a while everything is quiet, then the silence is broken by a rustling, shuffling sound. From a hole in a nearby earth bank appears a powerful chunky animal. It is a wombat, another of Australia's unusual pouched mammals.

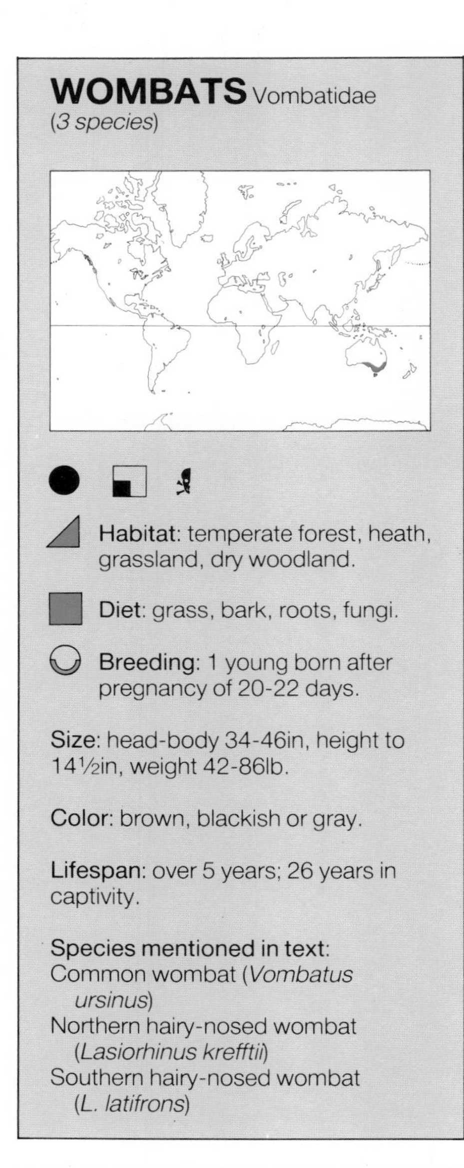

WOMBATS Vombatidae
(3 species)

Habitat: temperate forest, heath, grassland, dry woodland.

Diet: grass, bark, roots, fungi.

Breeding: 1 young born after pregnancy of 20-22 days.

Size: head-body 34-46in, height to 14½in, weight 42-86lb.

Color: brown, blackish or gray.

Lifespan: over 5 years; 26 years in captivity.

Species mentioned in text:
Common wombat (*Vombatus ursinus*)
Northern hairy-nosed wombat (*Lasiorhinus krefftii*)
Southern hairy-nosed wombat (*L. latifrons*)

◀Part of a wombat warren. In places the Common wombat is classed as a pest because it damages farmers' rabbit-proof fences.

▼To suit its burrowing lifestyle, the pouch of the female wombat opens towards the rear. The cub stays in the pouch until it is 6 to 7 months old.

The wombat is a perfectly designed earth-moving machine – nature's answer to the bulldozer. It is about 3ft long, with a heavy, thickset body and a short broad head. Its legs are short and powerful, and its feet are equipped with massive strong claws for digging.

Each animal lives in a maze of tunnels (a warren) up to 100ft in length, with several exits, side tunnels and resting chambers. Sometimes, neighboring warrens may overlap or even interconnect with each other, but the animals keep to themselves and lead mainly solitary lives except in the mating season.

WHO LIVES WHERE

There are three different species of wombat. The most widespread is the Common wombat, which lives in the eucalyptus woodlands of eastern and south-eastern Australia. The other two, called Hairy-nosed wombats, live in much drier parts, where food is even harder to find. The Southern hairy-nosed wombat inhabits dry salt-bush, scrub and savannah woodlands in south-central Australia and is still fairly common. Sadly, the Northern species is now very rare. Only about 20 animals are left, living in one small colony in the dry woodland of eastern Queensland.

SAVING VITAL ENERGY

Wombats live in areas where summer temperatures are high, where food is often scarce, and where there is very little drinking water. Understandably, their whole lifestyle is designed to avoid wasting precious energy and moisture. Most of the day is spent below ground, resting and keeping cool. To save energy, wombats allow their body temperature to fall and their heartbeat to slow while resting, then raise them again at night when it is time to go out in search of food.

Wombats feed mainly on coarse grass, roots and the bark of trees and shrubs. To cope with this tough woody diet they have a single pair of long front teeth which grow continuously, like those of rats and hamsters. Their digestive system is also designed to make the most of the low-quality food. Everything is digested very slowly, and all the valuable moisture is kept in the body. Hardly any is wasted. Their droppings are dry and they pass only tiny amounts of concentrated urine.

◀Hairy-nosed wombats are seasonal breeders, and usually produce a single cub in spring, when food is most plentiful. In times of drought they usually do not produce young.

▼The Common wombat is not a very aggressive animal. Individuals do fight occasionally, but the bites of the attacker seldom do much harm against the shaggy coat of the defender. The wombat's main natural enemy is the dingo.

HONEY POSSUM

As the Moon shines down on a patch of woodland in south-west Australia, a small shrew-like animal scrambles about among the enormous flower-heads of a *Banksia* tree. It is a Honey possum, and it has a special kind of partnership with the tree. As it takes its meal of sweet nectar and clambers over the slenderest of branches, the possum helps the tree by carrying its pollen from flower to flower.

▲The Honey possum's tail is longer than its head and body added together. The animal often uses it like an extra hand.

▼After a few minutes feeding, this Honey possum's face is dusted with the bright yellow pollen of a *Banksia* flower.

HONEY POSSUM
Tarsipes rostratus

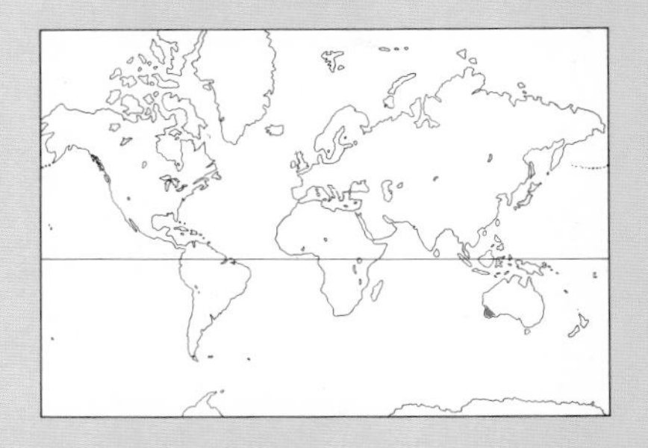

● ◪

◿ Habitat: heath, shrubland and open woodland with undergrowth.

■ Diet: nectar and pollen.

◎ Breeding: mainly in early summer. Litters of 2-4 born after pregnancy of about 28 days.

Size: head-body 2½-3½in, tail 3-4in; weight, male ¼-⅓ ounce, female ⅓-½ ounce.

Color: gray-brown above, pale cream below, with three stripes down back.

Lifespan: 1-2 years.

The Honey possum is found only in the heaths, shrublands and open woodlands of south-western Australia. Its name is rather misleading for it does not eat honey at all, but it certainly does like sweet food. It feeds mainly on nectar, and its body is highly specialized for this unusual way of life.

SPECIAL EQUIPMENT

The Honey possum has a long pointed snout for reaching deep inside the flowers it feeds on, and its tongue has a brush-like tip for lapping up the sugary nectar. It can run swiftly through the dense undergrowth, but it is also an expert climber. Its front and back feet are both designed for grasping, just like tiny hands, and this enables the animal to clamber about in the bushes, high among the thinnest twigs and flower stems. It can even grip with its tail, and often hangs upside-down by this extra "arm."

HITCH-HIKING POSSUMS

Honey possums are ready to mate when they are about 6 months old. There is no real courtship period. A male will simply follow a female until she is ready to mate, and once they have mated, the male goes off into the bush. Raising the family is left entirely to the female.

The baby possums are tiny when they are born, and they make their way at once into the mother's deep pouch. There they remain for 8 more weeks, feeding on milk from the four teats inside the pouch. By the time they are ready to leave the pouch for the first time, their eyes are open and they have a warm covering of fur. The whole breeding process is timed so that each litter is ready to leave the pouch when food is most plentiful, in the fall, spring or early summer.

At first the young are left in an old bird's nest or hollow tree while the mother feeds, but after a few days they climb on to her back and hitch a ride, although they soon become too heavy for her. At about 11 weeks they stop taking milk from the mother, and soon after this they set off to find a home of their own.

A PLACE TO LIVE

Outside the breeding season, Honey possums spend most of their time in separate home territories about one-and-a-half times the size of a soccer field. The territories overlap at the edges, but the animals seldom seem to fight with one another. However, things are different in the breeding season. Females with young spend their time in much smaller territories. These are strictly private, and strangers, especially males, are very quickly and aggressively chased away.

At present, survival of the Honey possum is not threatened. But being an animal that is found only in one small part of the world, it is likely quickly to become endangered as much of its habitat is destroyed. In the future it may be necessary to set up special reserves to protect this unusual little marsupial.

▼Honey possums are active mainly at night, and they rely on their sense of smell as well as their large eyes to find their food and to locate other animals.

GLOSSARY

Adaptation Features of an animal's body or life style that suit it to its environment.

Aggression Behavior in which one animal attacks or threatens another.

Aquatic Living for much, if not all the time in the water.

Arboreal Living for much, if not all, of the time in the trees. *Compare* terrestrial.

Arid Dry. Usually refers to land on the fringes of deserts.

Browsing The activity of a herbivore feeding on the leaves and branches of shrubs and trees.

Camouflage Color and patterns on an animal's coat that allow it to blend in with its surroundings.

Cheek pouches Loose folds of skin in the mouth, in which animals, such as hamsters, stuff food. They then take it to their food store.

Class The division of animal classification above Order.

Competition The contest between two or more species over such things as space and food.

Coniferous forest Forest of trees with cones and needle-like leaves, which usually bear leaves all year.

Conservation Preserving and protecting living things, their habitat and the environment in general.

Creche A kind of nursery den in which several mothers leave their young. The mara or Patagonian cavy does this.

Crepuscular Animal species that are active at dawn, dusk or both. Desert animals especially are usually crepuscular.

Deciduous forest Forest of trees that shed their leaves seasonally, usually in winter.

Den A shelter in which an animal or group of animals sleeps, hides, or gives birth to young.

Diet The food an animal eats.

Diurnal Active during the day.

Endangered species One whose numbers have dropped so low that it is in danger of becoming extinct.

Environment The surroundings, of a particular species, or the world about us in general.

Extinction The complete loss of a species, either locally or on the Earth.

Family The division of animal classification below Order and above Genus.

Feral Living in the wild; refers, for example, to animals that have escaped from captivity, such as the coypu in Europe.

Genus The division of animal classification below Family and above Species.

Gestation The period of pregnancy of an animal.

Grazing The activity of a herbivore feeding on vegetation close to the ground.

Guard hairs The long coarse hair of a furry animal such as the coypu, which protects the soft underfur.

Guinea pig The popular name for the rodents correctly called cavies, which belong to the family Caviidae. They are so called because they live in the wild in Guyana, in South America.

Habitat The kind of surroundings in which an animal lives.

Herbivore An animal that eats mainly plants.

Hibernation A winter period in which an animal is inactive.

Home range The area in which an animal usually lives and feeds.

Incisors The teeth at the front of the mouth, which in rodents can be highly developed. Mole-rats have long, protruding chisel-like incisors, which they use for digging.

Lagomorphs Animals belonging to the order Lagomorpha. They include rabbits, hares and pikas.

Leveret A young hare.

Lodge The shelter in which a beaver family spends the winter. All members of the family help build the lodge, which is begun or improved in late summer.

Mammals A class of animals whose females have mammary glands, which produce milk on which they feed their young.

Marsupials An order of mammals, whose females give birth to very under-developed young and then raise them (usually) in a pouch. The koala, wombat and the possums are marsupials.

Microtines Voles and lemmings, mammals that belong to the sub-family Microtinae.

Migration The long-distance movement of animals, usually seasonal, for the purposes of feeding or breeding.

Murine Member of the sub-family Murinae, the Old World rats and mice; it also means rat-like or mouse-like.

Myomorphs Mouse-like rodents, including rats and mice, lemmings, dormice and jerboas. They belong to the sub-order Myomorpha.

Myxomatosis A virus disease deadly to European rabbits. It was introduced to Australia in the 1950s to control the rabbit population, which had reached pest proportions. It was also introduced into Europe soon afterwards. Rabbits have now acquired some immunity to the virus.

Nocturnal Active during the night.

Nutria The fur of the coypu.

Omnivore An animal that has a varied diet, including both plants and animals.

Order The division of animal classification below Class and above Family.

Pampas The steppe grasslands of Argentina.

Phalangers Animals belonging to the family Phalangeridae, which includes the cuscuses and the brushtail possums.

Population cycle The fairly regular variation in the population of some species, such as voles and lemmings. Lemming populations reach large numbers every 4 years or so. Mass migrations follow in which hordes of the creatures are often killed when they try to make journeys they cannot perform.

Pouch A flap of skin, usually like a pocket, which covers the teats of female marsupials and in which the young are raised.

Prairie The steppe grasslands of North America.

Predator An animal that hunts live prey.

Pregnancy Period during which the young grows inside the body of a mammal.

Prehensile tail One that grips. The cuscuses and brushtail possums have a prehensile tail.

Race The division of animal classification below sub-species; it refers to animals that are very similar but have slightly different characteristics, eg Highland and Lowland gorillas.

Rain forest Tropical and sub-tropical forest which has plentiful rainfall all year round.

Rodents Animals belonging to the order Rodentia, including rats, mice, squirrels and cavies.

Savannah The tropical grassland of Africa, Central and South America and Australia.

Solitary Living alone for most of the time.

Species The division of animal classification below Genus; a group of animals of the same structure which can breed with one another.

Steppe The temperate grassy plains of Eurasia. Called "prairie" in North America.

Sub-species The division of animal classification below Species and above Race; typically the sub-species are separated geographically.

Taiga Coniferous forest of the far north, which is interspersed with boggy and rocky areas.

Tail-slapping One method beavers use to communicate. They slap their tail on the water as a warning to other beavers and as an attempt to frighten their enemies.

Terrestrial Spending most of the time on the ground. *Compare* arboreal.

Territory The area in which an animal or group of animals lives and defends against intruders.

Tundra The barren treeless land in the far north of Europe, Asia and North America. The vegetation includes low shrub, moss and lichen.

Ungulates Mammals that have hoofs. Most are large and live entirely on plant material.

Vegetarian An animal that lives entirely on plant material.

Warren The system of burrows in which rabbits live.

INDEX

Common names

Single page numbers indicate a major section of the main text. Double, hyphenated, numbers refer to major articles. **Bold numbers** refer to illustrations.

Scientific names

The first name of each double-barrel Latin name refers to the *Genus*, the second to the *species*. Single names not in *italic* refer to a family or sub-family and are cross-referenced to the Common name index.

Dipus sagitta (Feather-footed jerboa) 48
Dolichotis patagonum (mara) 54

Eliomys quercinus (Garden dormouse) 46
Ellobius fuscocapillus (Southern mole-vole) 39
Erethizon dorsatum (North American porcupine) 50,51,53
Erethizontidae *see* porcupines

Felovia vae (Felou gundi) 61

Geomyidae *see* gophers
Geomys bursarius (Plain's gopher) 21
Gerbillinae *see* gerbils
Gerbillurus paeba (South African pygmy gerbil) 45
Gerbillus
dasyurus (Wagner's gerbil) 44
gerbillus 45
Glaucomys volans (Southern flying squirrel)17
Gliridae *see* dormice
Glis glis (Edible dormouse) 46

Hesperomyinae *see* mice, New World; rats, New World
Heterocephalus glaber (Naked mole-rat) 64,65
Heterohyrax brucei (Bruce's yellow-spotted hyrax) 76
Hybomys univittatus (Peter's striped mouse) 33
Hydrochoerus hydrochaeris (capybara) 56-57
Hydromys chrysogaster (Australian water rat) 26
Hystricidae *see* porcupines
Hystrix species (crested porcupines) 50,53
Hystrix
africaeaustralis (Cape porcupine) 50,53
cristata (African porcupine) 50,51

Jaculus jaculus (Common desert jerboa) 48

Kerodon rupestris (Rock cavy) 54,55

Lagidium species (mountain viscachas) 62,63
Lagostomus maximus (Plains' viscacha) 62,63
Lasiorhinus
krefftii (Northern hairy-nosed wombat) 86
latifrons (Southern hairy-nosed wombat) 86
Lemmus lemmus (Norway lemming) 36,37,38,39
Leporidae *see* rabbits and hares
Lepus
americanus (Snowshoe hare) 68
californicus (Black-tailed jackrabbit) 68,71
europaeus (European hare) 66,70
timidus (Arctic hare) 71
Lophuromys sikapusi (Harsh-furred mouse) 35

Marmota
marmota (Alpine marmot) 14,18
monax (woodchuck) 16
Massoutiera mzabi (Mzab gundi) 61
Megalonychidae *see* sloths
Meriones
libycus (Libyan jird) 45
tamariscinus (Tamarisk gerbil) 45
unguiculatus (Mongolian gerbil) 44,45
Mesocricetus auratus (Golden hamster) 42
Micromys minutus (Harvest mouse) 31,32,34
Microtinae *see* voles and lemmings
Microtus
ochrogaster (Prairie vole) 38
pennsylvanicus (Meadow vole) 39
xanothognathus (Taiga vole) 39
Muridae *see* hamsters, lemmings, mice, rats, voles
Murinae *see* Mice, Old World; rats, Old World
Mus
minutoides (Pygmy mouse) 32,33
musculus (House mouse) 30,31,34,35
triton (Larger pygmy mouse) 33
Myocastor coypus (coypu) 58-59
Myosciurus pumilio (African pygmy squirrel) 14,17
Myospalax genus (zokor) 26

Nectomys species (South American water rats) 27
Neotoma micropus (North American woodrat) 25
Neotomodon alstoni (Volcano mouse) 33
Nesolagus netscheri (Sumatran hare) 67,71
Nesoryzomys genus (Galapagos rice rat) 24
Notiomys species (mole mice) 30,34
Notomys cervinus (Fawn hopping mouse) 35
Nyctomis sumichrasti (Central American vesper rat) 25

Ochotona
collaris (Collared pika) 72,73
macrotis (Large-eared pika) 72
princeps (North American pika) 73
pusilla (Steppe pika) 72
rufescens (Afghan pika) 72
Ochotonidae *see* pikas
Ondata zibethicus (muskrat) 36,39,41
Onychomys species (grasshopper mice) 33
Orthogeomys grandis (Large pocket gopher or taltuza) 20,21
Oryctolagus cuniculus (European rabbit) 66,67,70,71
Otomys species (African swamp rats) 26

Pappogeomys bulleri (Buller's gopher) 21
Paradipus ctenodactylus (Comb-toed jerboa) 48
Pectinator spekei (Speke's gundi) 60,61
Pedetes capensis (springhare) 22,23
Pentalagus furnessi (Amami rabbit) 67
Peromyscus species (deer mice) 32,33,34
Petaurista species (Giant flying squirrel) 16
Phalanger
lullulae (Woodlark Is. cuscus) 82
maculatus (Spotted cuscus) 81,82
orientalis (Gray cuscus) 81
Phalangeridae *see* phalangers
Phascolarctos cinereus (koala) 84-85
Phenacomys longicaudatus (Red-tree vole) 39
Phloeomys cumingi (Cuming's slender-tailed cloud rat) 24,26
Phodopus sungorus (Dzungarian hamster) 42
Phyllotis species (leaf-eared mice) 32
Platacanthomyinae *see* dormice
Platacanthomys lasiurus (Spiny dormouse) 46
Poelagus marjorita (Bunyoro rabbit) 67
Praomys
alleni (Climbing wood mouse) 33
natalensis (Multimammate rat) 28
Procavia johstoni (Johnston's hyrax) 76
Procaviidae *see* hyraxes
Pronolagus crassicaudatus (Greater red rock-hare) 66
Psammomys obesus (Fat sand rat) 45

Rattus
exulans (Polynesian rat) 28
muelleri (Muller's rat) 28
norvegicus (Norway rat) 24,25,26,27,28
rattus (Roof rat) 24,26,28
Ratufa indica (Indian giant squirrel) 17
Reithrodontomys humilis (American harvest mouse) 30
Rhabdomys pumilio (Four-striped grass mouse) 35
Rheomys species (water mice) 33

Rhinosciurus laticaudatus (Shrew-faced ground squirrel) 16,18
Rhipidomys genus (South American climbing rat) 24
Rhizomyinae *see* rats, root
Rhombomys opimus (Great gerbil) 45
Romerolagus diazi (Volcano rabbit) 66,71

Scapteromys tumidus (Argentinian water rat) 27
Sciuridae *see* squirrels
Sciurus
aberti (Abert or Tassel-eared squirrel) 17
carolinensis (Gray squirrel) 14,18,19
vulgaris (European red squirrel) 14,18
Seleviniidae *see* dormice
Sigmodon genus (cotton rat) 27
Spalax genus (Giant blind mole-rat) 26
Spermophilus beldingi (Belding's ground squirrel) 16,18
Sylvilagus species (cottontails) 68
Sylvilagus
floridanus (Eastern cottontail) 66
idahoensis (Pygmy rabbit) 66

Tachyoryctes splendens (East African root rat) 65
Tamias sibiricus (Siberian chipmunk) 18
Tamiasciurus hudsonicus (American red squirrel) 17
Tarsipes rostratus (Honey possum) 88-89
Thecarus species (Indonesian porcupines) 51
Thomomys bottae (Valley or Valley pocket gopher) 20,21
Trischosurus
caninus (Mountain brushtail possum) 81
vulpecula (common brushtail possum) 80,81

Tylomis genus (Central American climbing rat) 25

Vombatidae *see* wombats
Vombatus ursinus (Common wombat) 86,87

Wyulda squamicaudata (Scaly-tailed possum) 81,82

Xerus
erythropus (Western ground squirrel) 18
inaurus (Cape ground squirrel) 14

Zapodidae *see* mice, jumping
Zapus hudsonius (Meadow jumping mouse) 49
Zygogeomys trichopus (Michoacan gopher) 21

FURTHER READING

Alexander, R. McNeill (ed)(1986), *The Encyclopedia of Animal Biology*, Facts on File, New York
Berry, R.J. and Hallam, A. (eds)(1986), *The Encyclopedia of Animal Evolution*, Facts on File, New York
Corbet, G.B. and Hill, J.E. (1980), *A World List of Mammalian Species*, British Museum and Cornell University Press, London and Ithaca, NY
Delany, M.J. (1975), *The Rodents of Uganda*, British Museum (Natural History), London
Elton, C (1942), *Voles, Mice and Lemmings*, Oxford University Press, Oxford
de Graff, G. (1981), *The Rodents of Southern Africa*, Butterworth, Durban
Griffiths, M.E. (1978), *The Biology of Monotremes*, Academic Press, New York
Grzimek, B. (ed)(1972), *Grzimek's Animal Life Encyclopedia*, vols 10, 11, 12, Van Nostrand Reinhold, New York
Hall, E.R. and Kelson, K.R. (1959), *The Mammals of North America*, Ronald Press, New York
Harrison Matthews, L. (1969), *The Life of Mammals*, vols 1 and 2, Weidenfeld and Nicolson, London
Hunsaker II, D. (ed)(1977), *The Biology of Marsupials*, Academic Press, New York
Kingdon, J. (1971-82), *East African Mammals*, vols I-III, Academic Press, New York
Macdonald, D. (ed)(1984) *The Encyclopedia of Mammals*, Facts on File, New York
Moore, P. D. (ed)(1986), *The Encyclopedia of Animal Ecology*, Facts on File, New York
Morgan, L.H. (1868), *The American Beaver and his Works*, Burt Franklin, New York
Nowak, R.M. and Paradiso, J.L. (eds)(1983) *Walker's Mammals of the World* (4th edn) 2 vols, Johns Hopkins University Press, Baltimore and London
Rosevear, D.R. (1969) *The Rodents of West Africa*, British Museum (Natural History), London
Slater, P.J.B. (ed)(1986), *The Encyclopedia of Animal Behavior*, Facts on File, New York
Strahan, R. (ed), *The Complete Book of Australian Mammals*, Angus and Robertson, Sydney
Watts, C.H.S. and Aslin, H.J. (1981), *The Rodents of Australia*, Angus and Robertson, Sydney and London
Young, J.Z. (1975), *The Life of Mammals: their Anatomy and Physiology*, Oxford University Press, Oxford

ACKNOWLEDGMENTS

Picture credits

Key: *t* top *b* bottom *c* centre *l* left *r* right
Abbreviations: A Ardea. AH Andrew Henley. AN Nature, Agence Photographique. ANT Australasian Nature Transparencies. BC Bruce Coleman Ltd. FL Frank Lane Agency. J Jacana. NHPA Natural History Photographic Agency. OSF Oxford Scientific Films. PEP Planet Earth Pictures. SA Survival Anglia Ltd.

7 World Wildlife Fund/A.Purcell. 8 SA/J. & D. Bartlett. 12*t* BC 12*b* BC. 13*bl* FL/M. Newman, 13*br* Len Rue Jr. 14 SA/J. & D. Bartlett. 15 SA. 16*t* Bio-Tec Images, 16*b* Bio-Tec Images. 18 Bio-Tec Images. 19 M. Fogden. 20*b* A. 22 NHPA. 23 J. 25 R.W. Barbour. 26 AN. 28/29 Nature Photographers Ltd. 30 M. Fogden. 31 BC. 32*t* BC. 32*b* BC. 33 M. Fogden. 34 M. Fogden. 35 AH. 36 NHPA. 37*t* NHPA, 37*b* Naturfotografernas Bildbyra. 38 BC. 40/41 BC/Jane Burton. 42 NHPA. 43 BC. 45 Premaphotos. 47 AN. 48 AN. 49*t* R.W. Barbour, 49*b* OSF/J.A.L. Cooke. 50 A. Bannister. 51 BC. 52 Tony Morrison. 53 BC. 54 BC. 55 T. Owen-Edmunds. 56 D. Macdonald. 57*t* D. Macdonald, 57*b* D. Macdonald. 58 BC. 59*t* BC/Gordon Langsbury, 59*b* A. 60 W. George. 62 BC/Jane Burton. 63 Premaphotos. 65*t* J.U.M. Jarvis, 65*b* BC. 68/69 BC. 70/71 OSF/G. Bernard. 71 BC. 72 W. Ervin/Natural Imagery. 73*t* BC, 73*b* BC. 74 BC. 76 H. Hoeck. 77 H. Hoeck. 78/79 M. Fogden. 79 M. Fogden. 80*c* E. Beaton. 81*t* AH, 81*bl* ANT, 81*br* J. 82/83 A. 84*bl* AH, 84*br* AH. 85 AN. 86 BC. 87*t* BC/Francisco Erize, 87*b* ANT. 88 A. Smith. 89 BC/Jan Taylor.

Artwork credits

Key: *t* top *b* bottom *c* centre *l* left *r* right
Abbreviations: P B Priscilla Barrett.

6 Mick Saunders. 7*t* Jeane Colville, 7*b* Mick Saunders. 8 John Fuller. 9 Mick Saunders. 10/11 Milne & Stebbing Illustration. 13*t* Simon Driver. 17 PB. 18/19 PB. 20/21 PB. 22 PB. 24/25 PB. 27 PB. 30 PB. 31 Rob van Assen. 32 PB. 35 PB. 37 PB. 38/39 PB. 42 Rob van Assen. 44/45 PB. 46 PB. 48 Jeane Colville. 51 Denys Ovenden. 55 PB. 61 Denys Ovenden. 64 PB. 66/67 PB. 70 Jeane Colville. 75 Denys Ovenden. 86 Jeane Colville. 88 Denys Ovenden.